P9-CIU-465

The End of Doom

A Cato Institute Book

The End of Doom

Environmental Renewal in the Twenty-first Century

Ronald Bailey

THOMAS DUNNE BOOKS
ST. MARTIN'S PRESS ⚕ NEW YORK

THOMAS DUNNE BOOKS.
An imprint of St. Martin's Press.

www.thomas dunnebooks.com
www.stmartins.com

Significant portions of the present volume are based on my years of reporting on
these topics for *Reason* magazine.

Library of Congress Cataloging-in-Publication Data

Bailey, Ronald.
    The end of doom : environmental renewal in the twenty-first century / Ronald
Bailey.—First edition.
        p. cm.
    Includes bibliographical references and index.
    ISBN 978-1-250-05767-9 (hardcover)
    ISBN 978-1-4668-6144-2 (e-book)
    1. Environmental quality—Popular works.   2. Environmental protection—
Public opinion—Popular works.   3. Environmental disasters—Prevention—
Popular works.   I. Title.
    GE140.B3525 2015
    363.7—dc23

                                                                    2015013616

St. Martin's Press books may be purchased for educational, business, or promotional
use. For information on bulk purchases, please contact the Macmillan Corporate and
Premium Sales Department at 1-800-221-7945, extension 5442, or write to special
markets@macmillan.com.

First Edition: June 2015

10   9   8   7   6   5   4   3   2   1

For my love, Pamela Cole Friedman.
And for Madeleine, Anne, C.J., Lucy, Sarah, and Archie Lee,
with full confidence that they and their generation will
enjoy an abundant and amazing future.

CONTENTS

ACKNOWLEDGMENTS

FIRST, I MUST THANK MY COLLEAGUES AT *REASON* magazine and the Reason Foundation for their support of this project. Reason Foundation president David Nott especially encouraged me to write this book. I am particularly grateful to my brilliant editors at *Reason*, Nick Gillespie and Matt Welch, who provide an intellectually rich and fun venue in which to explore and report new ideas.

Of course, I owe a great debt of gratitude to my excellent editor at St. Martin's Press, Thomas Dunne, who published my first book on environmental issues way back in 1993. It is great to be working with him again.

I am most grateful for the generous and timely patronage provided by Ed Crane and David Boaz from the Cato Institute and Kim Dennis and Richard Tren from the Searle Freedom Trust.

I have benefited enormously from the insights and criticisms of

Terry Anderson from the Property Environment Research Center. Jerry Taylor merits special gratitude, since it was his idea that I write this book. Among the many others whom I must thank for championing me and my work are Matt Ridley, Daniel Benjamin, and Katherine Mangu-Ward. I also appreciate how patiently my longtime friends Steve and Leslie Frantz and Stuart and Brock Lending listened to and critiqued the ideas and data found in this book. I am forever grateful for meeting and learning from Norman Borlaug and Julian Simon. Their belief that human ingenuity will produce growing prosperity while safeguarding the natural world is being vindicated.

Finally, and most important, I am greatly indebted to the scores of researchers who shared their time, their scientific findings, their philosophical insights, and their hopes for the future of humanity with me.

INTRODUCTION

A LITTLE OVER TWO DECADES AGO, I WROTE a book, *Eco-Scam: The False Prophets of Ecological Apocalypse,* in which I looked closely at prevalent and generally accepted predictions of imminent planetary-scale environmental dooms. I analyzed the psychological appeal of doom, how predictions of disaster function as a political technique aimed at frightening people into handing over power to self-selected elites who want to enact drastic transformations in social and economic institutions. As I explained in my introduction twenty-two years ago, I was initially fascinated by these prophecies of global catastrophe because I had believed them.

Why? Because I had read and absorbed the messages of the classics of ideological environmentalism—*Silent Spring* by Rachel Carson, *The Population Bomb* by Paul Ehrlich, and *The Limits to Growth: A Report for the Club of Rome's Project on the Predicament of Mankind*—and believed that I and all of my fellow human beings were headed straight-away into a hopelessly bleak and hellish future. Synthetic chemicals were poisoning the natural world and human bodies. Overpopulation would soon outstrip the ability of farmers to grow enough food, and hundreds of millions would die in massive famines in the 1970s. And the world would shortly run out of oil and other nonrenewable resources, thus crashing modern civilization well before the year 2000.

Two decades after these and other dire predictions had been made, I noticed that we were still here and that civilization had not collapsed. Not only that, but people were actually living longer and healthier lives, famine had been held at bay, and the world was becoming more prosperous, not less. Naturally I was pleased that I was alive and that the dire environmentalist predictions had evidently not come true. At the time, I was a staff writer at *Forbes* magazine, where I conceived of a project in which I would go back and reread the classic prophecies of doom. I would write a series of articles after I interviewed their authors to see what they had to say about their prognostications. Obviously I couldn't talk with Rachel Carson, since she had died in 1964, but Ehrlich and the folks at the Massachusetts Institute of Technology (MIT) who had devised the computer program at the heart of the resource projections in *The Limits to Growth* were still around.

To make a long story short, I spoke with Ehrlich and he assured me that he had simply gotten his timing wrong. Globe-spanning famines would break out some time between 2000 and 2010. I also talked with Jay Forrester, the MIT systems dynamics professor who was the developer of the computer model used to make the forecasts in *The Limits to Growth*. He told me, "I think in retrospect that *Limits to Growth* overemphasized the material resources side." Well, yes. As I did more reporting on environmentalist doomsaying, it became increasingly apparent that the doomsters were not making scientific predictions, but instead were promoting a world view, an ideology—that is to say, *ideology* as properly defined as a body of doctrine, myth, belief, etc., that guides an individual, social movement, institution, class, or large group.

Chief among the doctrines in the ideology espoused by environmentalist doomsters is that nature is innocent and good and humanity evil. Environmentalist thinker Jeremy Rifkin explained the creed this way: "To end our long, self-imposed exile; to rejoin the community of life. This is the task before us. It will require that we renounce our drive for sovereignty over everything that lives; that we restore the rest of creation to a place of dignity and respect." He added, "Nature offers us the sublime resignation that goes with undifferentiated participation in the world around us." In other words, a secularized version of the myth of the Garden of Eden motivates many modern environmentalists.

Back in 1992, my book delved into the reasons why various prophecies of environmentalist doom had failed. Human beings are not like a herd of deer that simply starves to death when it overgrazes its meadow. Instead we seek out new ways to produce more food and do it ever more efficiently. I described how breakthroughs in plant breeding spawned the Green Revolution, which dramatically boosted global food production. When supplies of a resource run low, people use it more sparingly and find new sources and substitutes for it. Canadian environmental researcher Vaclav Smil calculates that back in 1920 in the United States it took about 10 ounces of materials to produce a dollar's worth of value, but that same value is now accomplished using only about 2.5 ounces, yielding a 75 percent decline in material intensity.

I discovered that it is almost always the case that wherever someone sees an environmental predicament in the world, it is a commons problem. The problem is occurring in an open-access commons, an area no one owns and for whose stewardship no one is responsible. The classic examples are fisheries. Frequently they are an open-access resource that is being overexploited. If a fisher leaves a fish in the water to spawn, the next guy will catch it and sell it. Thus no individual fisher has the incentive to protect the health and productivity of the fishery. It's a race to the bottom, with both fish and fishers losing out. Similarly, pollutants are pumped into rivers and into the air and tropical forests are chopped down because all too often anyone can use those resources without paying for the costs of the harm they cause.

One such commons problem I considered in my 1992 book was the "ozone hole" over Antarctica that was produced by chlorofluorocarbon (CFC) refrigerants floating up into the stratosphere. Ozone consists of three oxygen atoms linked together. At ground level ozone is a pollutant, but it functions as a sunscreen in the upper atmosphere, protecting living things against damaging ultraviolet light. The CFC pollutants were reacting in the cold stratosphere over Antarctica to erode the ozone layer there. I cited research that suggested that the apocalyptic assessments of many alarmists were unwarranted, but did agree that an international treaty was needed to phase out and replace the harmful refrigerants. The good news is that French researchers in 2013 reported that the Antarctic ozone shows a significant positive trend toward its recovery.

Another two decades have now passed and environmental gloom continues to be widely preached. For example, in 2013, Earth Policy Institute founder Lester Brown asserted, "The world is in transition from an era of food abundance to one of scarcity." During a lecture in 2013, the author of *The Population Bomb*, Paul Ehrlich, rhetorically asked, "What are the chances a collapse of civilization can be avoided?" Ehrlich answered himself: 10 percent. The German think tank Energy Watch Group declared that global oil production had peaked in 2006 and that supplies would be cut in half by 2030, triggering the "meltdown of society." Similarly, in 2010, a senior fellow at the Post Carbon Institute, Richard Heinberg, stated: "The world is at, nearing, or past the points of peak production of a number of critical nonrenewable resources—including oil, natural gas, and coal, as well as many economically important minerals ranging from antimony to zinc." And in 2014, the Center for Biological Diversity warned, "It could be a scary future indeed, with as many as 30 to 50 percent of all species possibly heading toward extinction by mid-century." In 2013, the United Nations Intergovernmental Panel on Climate Change affirmed, "Warming of the climate system is unequivocal, and since the 1950s, many of the observed changes are unprecedented over decades to millennia."

Globe-spanning famines have yet to occur; instead, average life expectancy has increased and a higher percentage of people are enjoying the benefits of modern technology than ever before. World population growth is slowing; and the number of human beings on the planet will likely peak and begin to fall toward the middle of this century. Pollution levels are falling in rich countries and will begin to drop in poor countries as they become wealthier. Similarly, forests are regrowing in many parts of the world. In fact, except in the cases of Indonesia and Brazil, globally the forests of the world have increased by about 2 percent since 1990.

As this present volume will make clear, I have changed my mind since 1992 about how big a problem man-made global warming might become. On the other hand, more than twenty years of reporting on United Nations climate change negotiations has convinced me that the ongoing attempt to hammer out a global treaty imposing hard, legally binding limits on the emissions of carbon dioxide and other gases that contribute to man-made warming is doomed to failure. Instead, I show how human ingenuity will likely solve this problem just as it did those that provoked earlier (and incorrect) predictions of environmental apocalypse. In this case, promising research suggests that it will be possible to lower the price of clean energy below that of fossil fuels in the next four decades and thus to lessen concerns about future disruptive climate change.

When I presented my book proposal to my editor, Thomas Dunne, at St. Martin's Press back in 1992, he actually told me: "Ron, we'll publish your book and we'll both make some money. But I want to tell you that if you'd brought me a book predicting the end of the world, I could have made you a rich man." Human beings do have a psychological bias toward believing bad news and discounting good news. But besides that, the sciences surrounding environmental issues have been politicized from top to bottom.

As the researchers at the Yale Cultural Cognition Project have shown time and again, what people believe about scientific issues is chiefly determined by their cultural values. They use a theory of cultural commitments devised by University of California at Berkeley political scientist Aaron Wildavsky that "holds that individuals can be expected to form perceptions of risk that reflect and reinforce values that they share with others."

Based on Wildavsky's typology, the Yale researchers divvy up Americans into four cultural groups: Individualists, Communitarians, Hierarchicalists, and Egalitarians. In general, Hierarchical folks prefer a social order where people have clearly defined roles and lines of authority. Egalitarians want to reduce racial, gender, and income inequalities. Individualists expect people to succeed or fail on their own, while Communitarians believe that society is obligated to take care of everyone.

The Yale researchers report that people whose values are located in Individualist/Hierarchy space "can be expected to be skeptical of claims of environmental and technological risks. Such people, according to the theory, intuitively perceive that widespread acceptance of such claims would license restrictions on commerce and industry, forms of behavior that Hierarchical/Individualists value." On the other hand, Egalitarian/Communitarians "tend to be morally suspicious of commerce and industry, which they see as the source of unjust disparities in wealth and power. They therefore find it congenial, the theory posits, to see those forms of behavior as dangerous and thus worthy of restriction." According to this view, then, Egalitarian/Communitarians would be more worried about all sorts of alleged environmental risks than would be Hierarchical/Individualists.

As the Yale Cultural Cognition Project researchers depressingly show in their numerous studies, people are adept at seeking out information that confirms their values while determinedly ignoring data that challenges them. While I have tried hard to avoid succumbing to confirmation bias in what I report, I suspect that it will be apparent when you read this book that I am, culturally speaking, an Individualist. So be skeptical on that account, but also be wary of your own susceptibility to confirmation bias.

In my introduction to my 1992 book, I concluded:

This book demonstrates the reality of human progress, and I hope it will thereby help restore the next generation's belief in its future. I do not counsel mindless boosterism or Panglossian optimism. The world faces some real problems, but those problems do not portend the end of the world. And yes, there are sometimes unintended consequences to human actions. However, history shows that our energy and creativity will surmount whatever difficulties we encounter. Life and progress will always be a struggle and humanity will never lack for new challenges, but as the last fifty years of solid achievement show, there is nothing out there that we can't handle.

Add twenty-two years to that.

I aim in this new book to again remind the public, the media, and policymakers that the foretellers of ruin have consistently been wrong, whereas the advocates of human resourcefulness have nearly always been right. So instead of ecological collapse, I predict that humanity can look hopefully forward in the twenty-first century to an age of environmental renewal.

The End of Doom

Peak Population?

"A MAN YOUR AGE WITH NO CHILDREN?" blurted out my flabber-gasted Johannesburg taxi driver. I was then forty-eight years old and taking a long cab ride from Sandton, where the UN's 2002 World Summit on Sustainable Development had convened, to a Soweto township venue in South Africa. As is my usual practice, I had gotten into the front seat to make conversation.

After we'd gone through the preliminaries about what had brought me to Johannesburg, my very hospitable driver, who looked to be in his fifties, asked me if I was married. "Yes," I replied. "How many children do you have?" he amiably asked. "None," I replied. My driver nearly veered off the road in shock. "Are there problems?" he carefully inquired. "No," I replied, "my wife and I decided not to have children." "Who is going to support you in your old age?" he wondered. I elected not to try to explain to him about retirement funds and Social Security and instead turned the questions around, asking him, "Well, how many children do you have?"

"Six," he replied with evident satisfaction. "Why so many?" I asked. "Because two of them are going to be rotters and just leave you," he genially explained. "Two others will support you when you get old. And you need two younger ones at home to fetch you beers from the fridge after work."

"How many children did your father have?" I asked. "I am one of twelve," my driver replied, adding that he had grown up on a farm in the northern part of the country. Then I asked, "How many children do your kids have now? "None," he replied with a hint of a frown. "City life is so expensive, a couple are still finishing up school, and good jobs are hard to find," he explained.

That's the demographic transition right there, I thought to myself.

Population researchers define the demographic transition as the change in the human condition from high mortality and high fertility to low mortality and low fertility. Initially, both birthrates and death rates are high and natural population growth is low. With the advent of modern medicine and sanitation, mortality rates fall and fertility remains high, producing a rapidly growing population. Eventually fertility rates also fall, leading to a reduction in the rate of population growth. "Population increases not because people start breeding like rabbits, but because they stop dying like flies," explains American Enterprise Institute demographer Nicholas Eberstadt.

In premodern societies, average life expectancy was under forty years, nearly a third of children died before reaching age five, child labor was vital to mostly rural families, and few women had access to education or contraception. Global average life expectancy is now over seventy years, only one in twenty children die before their fifth birthdays, urbanized child-rearing is costly, and many more women are educated and have access to contraception.

As a result of these trends, women in 1970 globally averaged 4.7 children over the courses of their lives, and that has fallen to 2.45 children in 2013. A population becomes stable when as many people are born as die. This occurs when the total fertility rate is approximately 2.1 children per woman; the extra tenth of a child takes into account pre-reproduction deaths, infertility, and people like my wife and me who choose not to have children.

Leading demographers expect that as the twenty-first century unfolds, women across the globe will be giving birth to fewer and fewer children. The upshot will be slowing population growth and eventually a reversal of trend in which world population begins to shrink. According to a long line of environmentalist doomsayers, this was not supposed to happen.

The Population Bomb

"The battle to feed all of humanity is over. In the 1970s the world will undergo famines—hundreds of millions of people are going to starve to death in spite of any crash programs embarked upon now," predicted Stanford University biologist Paul Ehrlich in his 1968 bestseller, *The Population Bomb*. Ehrlich was not a lone voice proclaiming the advent of imminent massive famines. A year earlier in their bestselling book, *Famine 1975! America's Decision: Who Will Survive?*, William and Paul Paddock warned, "By 1975 a disaster of unprecedented magnitude will face the world. Famines, greater than any in history, will ravage the undeveloped nations. A swelling population is blotting up the earth's food." They confidently added, "Our technology will be unable to increase food production in time to avert the deaths of tens of millions people by starvation."

In his famous 1968 essay "The Tragedy of the Commons," published in the journal *Science*, ecologist Garrett Hardin flatly declared, "The freedom to breed is intolerable." To illustrate the harms of the freedom to breed, he conjures up the arresting example of a pasture open to all people in a village. Each herdsman, seeking to maximize his individual gain, puts as many cattle on the pasture as possible, leading eventually to its destruction from overgrazing. "Ruin is the destination toward which all men rush, each pursuing his own best interest in a society that believes in the freedom of the commons," wrote Hardin. "Freedom in a commons brings ruin to us all." According to Hardin, ceaselessly breeding human beings treat the Earth like a village commons and would soon "overgraze" the planet. Hardin thus concluded that "the inherent logic of the commons remorselessly generates tragedy."

Another prominent prognosticator of population doom is Lester Brown, the founder of the Worldwatch Institute and the Earth Policy Institute. Back in 1963, when Brown was a young bureaucrat in the US Department of Agriculture, he declared, "The food problem emerging in the less-developing regions may be one of the most nearly insoluble problems facing man over the next few decades." In 1967, Brown explained, "As the non-recurring sources of [agricultural] productivity are exhausted . . . the rate of increase in yield per acre begins to slow." In 1974, Brown maintained that farmers "can no longer keep up with rising demand; thus the outlook is for chronic scarcities and rising prices." In 1989, Brown stated that "global food insecurity is increasing," and further claimed that "the slim excess of growth in food production over population is narrowing." Brown contended that "population growth is exceeding the farmer's ability to keep up," concluding that "our oldest enemy, hunger, is again at the door." In 1995, Brown starkly warned, "Humanity's greatest challenge may soon be just making it to the next harvest." In 1996, Brown again proclaimed, "Food scarcity will be the defining issue of the new era now unfolding." In a 2012 *Scientific American* article, Brown asked, "Could food shortages bring down civilization?" Not surprisingly, Brown's answer was an emphatic yes. Given his past record, he astonishingly claimed that for years he has "resisted the idea that food shortages could bring down not only individual governments but also our global civilization." Now, however, Brown said, "I can no longer ignore that risk." Also in 2013, Brown once again declared, "The world is in transition from an era of food abundance to one of scarcity."

The population doomsters offered stark plans to handle the impending global famines. The Paddock brothers advised a form of triage, in which the United States would pick countries worthy of food aid and leave tens of millions of people in India, Haiti, and Egypt to starve to death. In *The Population Bomb* Ehrlich compared humanity to a growing cancer on the Earth. "A cancer is an uncontrolled multiplication of cells; the population explosion is an uncontrolled multiplication of people," wrote Ehrlich. What must be done? "We must shift our efforts from treatment of the symptoms to the cutting out of the cancer. The operation will demand many apparently brutal and heartless decisions." What sorts of heartless decisions? The November 25, 1969, *The New York Times* reported, "Dr. Paul Ehrlich says the U.S. might have to resort to addition of temporary sterility drugs to food shipped to foreign countries or their water supply with limited distribution of antidote chemicals, perhaps by lottery."

That was then, but what about now?

Though it is tragically true that over the decades tens of millions have died of the effects of malnutrition, the world-spanning massive famines predicted by Ehrlich and the Paddocks did not come to pass. Instead, world population since 1968 has essentially doubled from 3.6 billion to 7.2 billion today. While the overpopulation dirge has become somewhat muted as a result of their massive predictive failure, many of the more radical environmentalist ideologues still sing the same old Malthusian song.

Malthusianism Forever?

Modern promoters of imminent population doom are the intellectual disciples of the eighteenth-century economist Reverend Thomas Robert Malthus. In the notorious first edition of his *An Essay on the Principle of Population*, Malthus claimed that human numbers would always outrun the amount of food available to feed people. Malthus advanced two propositions that he regarded as completely self-evident. First, that "food is necessary for the existence of man," and second, that "the passion between the sexes is necessary and will remain nearly in its present state." Based on these propositions, Malthus concluded that "the power of population is indefinitely greater than the power in the earth to produce subsistence for man. Population, when unchecked, increases in a geometrical ratio. Subsistence increases only in an arithmetical ratio. A slight acquaintance with numbers will show the immensity of the first power in comparison with the second." In other words, Malthus was arguing that population doubled at an exponential rate of 2, 4, 8, 16, 32, and so forth, whereas food production increased additively, rising one unit at a time, like 2, 3, 4, 5, 6, 7, 8, and so forth. Malthus additionally asserted that "population does invariably increase where there are the means of subsistence." Malthus therefore dismally concluded that some portion of humanity must forever be starving to death.

According to Malthus, there are two kinds of checks on population, preventive and positive. Preventive checks, those that prevent births, include abortion, infanticide, and prostitution; positive checks include war, pestilence, and famine. In later editions of his essay, Malthus added a third check that he called "moral restraint," which includes voluntary celibacy, late marriage, and the like. Moral restraint is basically just a milder version of the earlier preventive check.

If all else failed to keep human numbers under control, Malthus chillingly reckoned:

Famine seems to be the last, the most dreadful resource of nature. The power of population is so superior to the power in the earth to produce subsistence for man, that premature death must in some shape or other visit the human race. The vices of mankind are active and able ministers of depopulation. They are the precursors in the great army of destruction, and often finish the dreadful work themselves. But should they fail in this war of extermination, sickly seasons, epidemics, pestilence, and plague, advance in terrific array, and sweep off their thousands and ten thousands. Should success be still incomplete, gigantic inevitable famine stalks in the rear, and with one mighty blow, levels the population with the food of the world.

Reading Malthus in 1838 was a eureka moment for the founding father of modern biology, Charles Darwin, who declared in his autobiography, "I had at last got a theory by which to work." Darwin realized that Malthus's thesis applied to the natural world, since plants and animals produce far more offspring than there are food, nutrients, and space to support them. Consequently, Darwin noted, "It at once struck me that under these circumstances favourable variations would tend to be preserved, and unfavourable ones to be destroyed. The results of this would be the formation of a new species." This insight formed the basis for one of the most important modern scientific theories, the theory of biological evolution by means of natural selection.

Ever since, biologists have been entranced by the idea that if Malthusianism can explain the operation of the natural world, it should also explain the functioning of human societies. Are we not just complicated animals? Shouldn't this biological insight apply to us, too?

The Neo-Malthusians

The most prominent among the neo-Malthusians is Paul Ehrlich. Despite his utter failure as a prophet, Ehrlich continues to preach that overpopulation is humanity's biggest problem. "The human predicament is driven by overpopulation, overconsumption of natural resources, and the use of unnecessarily environmentally damaging technologies and socio-economic-political arrangements to service Homo sapiens' aggregate consumption," wrote Ehrlich and his wife, Anne, in the March 2013 issue of the *Proceedings of the Royal Society B*. During a May 2013 conference at the University of Vermont, Ehrlich asked, "What are the chances a collapse of civilization can be avoided?" His answer was 10 percent.

Even now the Ehrlichs are far from alone in propagating forecasts of overpopulation doom. "The world faces a serious overpopulation problem," asserted Cornell University researcher David Pimentel in his 2011 article "World Overpopulation." "The world's biggest problem?" asks a 2011 op-ed by researchers Mary Ellen Harte and Anne Ehrlich in the *Los Angeles Times*. "Too many people," they answer. "We are a plague upon the earth," declared nature documentarian Sir David Attenborough. "Either we limit our population growth or the natural world will do it for us." Attenborough expressed these dour sentiments in *The Telegraph* in January 2013.

In his 2013 rant *Ten Billion*, Microsoft Research computer scientist Stephen Emmott argued that humanity's growing population constitutes "an unprecedented planetary emergency." Emmott asserts, "As the population continues to grow, our problems will increase. And this means that every way we look at it, a planet of ten billion people is likely to be a nightmare." In the somewhat more hopeful 2013 book *Countdown: Our Last, Best Hope for a Future on Earth?*, journalist Alan Weisman declares that "this will likely be the century that determines what the optimal human population is for our planet." We can choose to limit population growth, argues Weisman, "or nature will do it for us, in the form of famines, thirst, climate chaos, crashing ecosystems, opportunistic disease, and wars over dwindling resources that finally cut us down to size."

More Food Equals More Kids?

In fact, the chief goal of most species is to turn food into offspring: the more food, the more offspring. "To ecologists who study animals, food and population often seem like sides of the same coin," wrote Paul and Anne Ehrlich in 1990. "If too many animals are devouring it, the food supply declines; too little food, the supply of animals declines." They further asserted, "*Homo sapiens* is no exception to that rule, and at the moment it seems likely that food will be our limiting resource." By *limiting*, they meant starvation.

Neo-Malthusians like the Ehrlichs, Pimentel, and Emmott cannot let go of the simple but clearly wrong idea that human beings are no different than a herd of deer when it comes to reproduction. For example, in an article called "Human Carrying Capacity Is Determined by Food Availability," in the November 2003 issue of the journal *Population and Environment,* Duke University researcher Russell Hopfenberg wrote: "The problem of human population growth can be feasibly addressed only if it is recognized that increases in the population of the human species, like increases in the population of all other species, is a function of increases in food availability." More food means more kids.

It is true that as food supplies have increased, so have human numbers. But Hopfenberg and other neo-Malthusians are overlooking some crucial data. The countries with the greatest food security are also the countries that are experiencing below replacement fertility. High fertility does not correlate with improved food availability. Consider that of the thirty-four nations that are members of the rich country club the Organisation for Economic Co-operation and Development, only Mexico has an above replacement rate fertility of 2.22 children. All the rest are at or well below the replacement rate. More generally, as food security has increased around the world, instead of increasing as Hopfenberg's neo-Malthusian theory would suggest, global average fertility rates have dropped from around 5 children per woman in 1960 to 2.45 today.

Instead, the highest fertility rates occur in countries where food insecurity is greatest. The International Food Policy Research Institute's *2013 Global Hunger Index* takes into account undernourishment rates, percentage of underweight children, and child mortality rates in various countries. Of the nineteen countries where the level of hunger was rated as alarming or extremely alarming, fifteen had total fertility rates higher than 4.5 children per woman. It is notable that Niger's total fertility rate of 7.6 children per woman is the highest in world. As we shall see, demographers have developed persuasive explanations for why people in countries suffering from food insecurity choose to have more children.

So how did humanity avoid the massive famines so confidently predicted by environmentalist millenarians like Ehrlich, Brown, and the Paddocks? Unlike deer that starve when their food runs out, people work to increase supplies. As it turns out, food plants and animals are populations, too, and can be, contrary to Malthus, increased at exponential rates.

Norman Borlaug and the Green Revolution

Norman Borlaug is the man who saved more human lives than anyone else in history. Borlaug was the father of the Green Revolution, the dramatic improvement in agricultural productivity that swept the globe in the 1960s. For spearheading this achievement, he was awarded the Nobel Peace Prize in 1970. One of the great privileges of my life was getting to meet and talk with Borlaug many times. He died in 2009 at age ninety-five.

Borlaug grew up on a small farm in Iowa and graduated from the University of Minnesota, where he studied forestry and plant pathology, in the 1930s. In 1944, the Rockefeller Foundation invited him to work on a project to boost wheat production in Mexico. At that time, Mexico could not feed itself and was importing half of its wheat supplies. Backed by $100,000 in annual funding from the foundation, Borlaug and his colleagues succeeded brilliantly in boosting the productivity of poor Mexican farmers. They did this by breeding new, highly productive dwarf wheat varieties that enabled Mexico to become self-sufficient in grains by 1956. By 1965, Mexican wheat yields had risen 400 percent over their levels in 1950.

In 1952, the Rockefeller Foundation began funding a similar effort to boost the productivity of poor farmers in India. In the mid-1960s, India was importing grains to avert looming famines. The dwarf wheat varieties developed by Borlaug and his colleagues were again decisive in winning the battle against hunger on the subcontinent.

In the late 1960s, as noted earlier, predictions of imminent global famines in which billions would perish were widespread. Recall that chief among the doomsters was *The Population Bomb* author Paul Ehrlich, who, as we've seen, predicted that in the 1970s "millions of people will starve to death in spite of any crash programs embarked upon now." Ehrlich also declared, "I have yet to meet anyone familiar with the situation who thinks India will be self-sufficient in food by 1971." And he further insisted that "India couldn't possibly feed two hundred million more people by 1980."

As we now know, Borlaug and his team were already engaged in exactly the kind of crash program that Ehrlich declared wouldn't work. Their dwarf wheat varieties resisted a wide spectrum of plant pests and diseases and produced two to three times more grain than the traditional varieties. In 1965, they launched a massive campaign to ship the miracle wheat to Pakistan and India and teach local farmers how to cultivate it properly. Soon after Borlaug's success with wheat, his colleagues at the Consultative Group on International Agricultural Research working at the International Rice Research Institute in the Philippines developed high-yield rice varieties that quickly spread the Green Revolution through most of Asia. By 1968, when Ehrlich's book appeared, the US Agency for International Development was already hailing Borlaug's achievement as a Green Revolution.

Borlaug's achievements were not confined to the laboratory and fields. He insisted that governments pay poor farmers world prices for their grain. At the time, many developing nations—eager to supply cheap food to their urban citizens, who might otherwise rebel—required their farmers to sell into a government concession that paid them less than half of the world market price for their agricultural products. The result, predictably, was hoarding and underproduction. Using his hardwon prestige as a kind of platform, Borlaug persuaded the governments of Pakistan and India to drop such self-defeating policies. Fair prices and high doses of fertilizer combined with new grains changed everything. By 1968 Pakistan was self-sufficient in wheat, and by 1974 India was self-sufficient in all cereals.

Instead of cheering the successes of the Green Revolution, Ehrlich doubled down on his predictions of imminent global collapse. In a 1969 article, "Eco-Catastrophe," in *Ramparts* magazine, he excoriated people "lacking the expertise to see through the Green Revolution drivel" for failing to realize that by "the early 1970s, the 'Green Revolution' was more talk than substance." Ehrlich derided officials in the US Department of Agriculture and Agency for International Development for supposedly "rav[ing] about the approaching transformation of agriculture in the underdeveloped countries (UDCs)." He continued, "Most historians agree that a combination of utter ignorance of ecology, a desire to justify past errors, and pressure from agroindustry" was behind the Green Revolution propaganda campaign. In his dire scenario, famines would soon break out first in India and Pakistan, but soon spread to "Indonesia, the Philippines, Malawi, the Congo, Egypt, Colombia, Ecuador, Honduras, the Dominican Republic, and Mexico." Ehrlich's prophecy of famine ends by proclaiming, "Everywhere hard realities destroyed the illusion of the Green Revolution."

The famines didn't happen. In Pakistan, wheat yields rose from 4.6 million tons in 1965 to 8.4 million in 1970. In India, wheat yields rose from 12.3 million tons to 20 million. And the yields continue to increase. The US Department of Agriculture is projecting Pakistan's 2014 wheat harvest at 24.5 million tons and India's at a record 96 million tons. Since Ehrlich's dire predictions in 1968, India's population has risen from 500 million to 1.2 billion and its economy has grown tenfold. Concurrently, its wheat production has also increased nearly fivefold. Both Pakistan and India export grain today. India is expected to export 18 million tons of grain in 2014.

Contrary to Ehrlich's bold pronouncements, hundreds of millions didn't die in massive famines. India fed far more than 200 million more people, and by 1971 it was close enough to self-sufficiency in food production that Ehrlich discreetly omitted his prediction about that from later editions of *The Population Bomb*. The last four decades have seen a "progress explosion" that has handily outmatched any "population explosion."

The Food and Agriculture Organization's global food production index (2004−2006 = 100) rose from 36 to 117 between 1961 and 2012. That means that over the past fifty or so years, world food production has more than tripled. In the meantime, world population increased from just over 3 billion in 1961 to 7.1 billion people in 2012, and the amount of food per person increased by about a third. The FAO further reports that between 1961 and 2009 (the latest figures available), global per capita annual consumption of cereals increased from 282 to 327 pounds, vegetables from 140 to 290 pounds, and meat from 51 to 92 pounds.

According to a 2008 World Bank report, as a result of the increase in food supplies, per capita consumption in developing countries rose from an average of 2,100 calories per day in 1970 to almost 2,700 calories today. That report further observes that "the proportion of people suffering from hunger has fallen by half since the 1960s, from more than one in three to one in six, even as [the] world's population has doubled." While progress has been made, some 870 million people are still undernourished.

Occasional local famines in poor countries caused by armed conflicts or political mischief do still occur. But food is more abundant today than ever before in history, due in large part to the work of Borlaug and his colleagues, and no thanks to neo-Malthusian false prophets like Paul Ehrlich and Lester Brown.

Can We Feed the World in 2050?

Looking to the future, what is likely to happen? The International Food Policy Research Institute projects that farmers will have to produce about 70 percent more food than they do today in order to provide the projected population in 2050 with a nutritionally satisfactory diet.

The journal *Philosophical Transactions of the Royal Society B*

(Biological Sciences) devoted its September 27, 2010, issue to analyzing the issue of global food security through 2050. In one of the specially commissioned research articles, it is projected that world population will reach around 9 billion by 2050, and that in the second half of the twenty-first century, "population stabilization and the onset of a decline are likely." Can the world's farmers be reasonably expected to provide enough food for 9 billion people by 2050?

Two other articles in the special Royal Society issue on global food security conclude yes. A review of the relevant scientific literature led by Keith Jaggard from Rothamsted Research looks at the effects of climate change, CO_2 increases, ozone pollution, higher average temperatures, and other factors on future crop production. Jaggard and his colleagues conclude: "So long as plant breeding efforts are not hampered and modern agricultural technology continues to be available to farmers, it should be possible to produce yield increases that are large enough to meet some of the predictions of world food needs, even without having to devote more land to arable agriculture."

Applying modern agricultural technologies more widely would also go a long way toward boosting yields. In his 1997 article "How Much Land Can Ten Billion People Spare for Nature?," published by the National Academy of Engineering, agronomist Paul Waggoner argued that "if during the next sixty to seventy years the world farmer reaches the average yield of today's U.S. corn grower, the 10 billion will need only half of today's cropland while they eat today's American calories."

University of Minnesota biologist Ronald Phillips points out that India produces 31 bushels of corn per acre now, which is at the same point US yields were in the 1930s. Similarly, South Africa produces 40 bushels (US 1940s yields); Brazil 58 bushels (US 1950s yields); China 85 bushels (US 1960s yields). Today's modern biotech hybrids regularly produce more than 160 bushels of corn per acre in the Midwest. For what it's worth, the corporate agriculture giant Monsanto is aiming to double yields on soybeans and cotton by 2030. Whether or not specific countries will be able to feed themselves has less to do with their population growth than it does with whether they adopt policies that retard their economic growth.

An Overpopulated Nightmare?

Will the twenty-first century be an overpopulated nightmare, as Stephen Emmott asserts? There are good reasons to doubt it. First, let's take a look at the latest population projections by the United Nations. Every two years the United Nations Population Fund issues estimates for future population. In their latest report, *World Population Prospects: The 2012 Revision,* demographers at the United Nations boosted projected world population numbers by 600 million. The UN experts offer low-variant, middle-variant, high-variant, and constant fertility trends. The new estimates of the middle-variant projections of future population in 2050 increased from 9 billion in the UN's *2010 Revision* to 9.6 billion and from 10 billion to 10.9 billion by 2100.

The UN's middle-variant projection is generally taken to be the most likely path of future population growth. The difference between the low- and the high-variant projections is basically one child. In the new low-variant projection, world population would reach 8.3 billion by 2050, whereas the high-variant projection would result in a population of 10.9 billion by then. As the report explains, "Thus, a constant difference of only half a child above or below the medium variant would result in a global population in 2050 of around 1.3 billion more or less compared to the medium variant of 9.6 billion."

UN estimates are not universally accepted. Many other demographers believe that the new UN projections are too high. For example, in a 2013 study, Félix-Fernando Muñoz and Julio A. Gonzalo, researchers associated with the Spanish Foundation for Science and Technology, find that past population growth has generally followed the UN's low-variant trend. Using sophisticated statistical techniques, the two calculate that future population growth will most likely continue to track the UN's low-variant trends. "Overpopulation was a spectre in the 1960s and 70s but historically the UN's low fertility variant forecasts have been fulfilled," noted Muñoz. If the Spanish researchers are right, world population will top out at between 8 and 9 billion by mid-century and thereafter begin declining.

In 2001, International Institute for Applied Systems Analysis (IIASA) demographer Wolfgang Lutz and his colleagues published "The End of World Population Growth" in the journal *Nature*. Lutz and his fellow researchers calculated that "there is around an 85 per cent chance that the world's population will stop growing before the end of the century. There is a 60 per cent probability that the world's population will not exceed 10 billion people before 2100, and around a 15 per cent probability that the world's population at the end of the century will be lower than it is today." In a 2013 study in *Demographic Research*, the IIASA researchers noted that "most existing world population projections agree that we are likely to see the end of world population growth (with a peak population of between eight and ten billion) during the second half of this century." In another 2013 study for the United Nations, the IIASA demographers project that world population will most likely peak around 2070 at 9.4 billion and fall back below 9 billion by 2100.

In a September 2013 Deutsche Bank report, demographer Sanjeev Sanyal argued that the latest UN population projections are way too high and that world population will likely peak at 8.7 billion around 2050 and then begin falling. Sanyal noted that in recent decades total fertility rates have fallen much more sharply than predicted in countries like China, India, Iran, and Bangladesh. He makes the case that rates are at the brink of similarly steep declines in current high-fertility countries such as Nigeria and Pakistan. As a consequence, Sanyal argues that "the world's overall fertility rate will fall to replacement rate by 2025. In other words, reproductively speaking, our species will no longer be expanding—a major turning point in history." Thus, Sanyal and his colleagues predict, "World population will peak around 2055 at 8.7 billion and will then decline to 8 billion by 2100. In other words, our forecasts suggest that world population will peak at least half a century sooner than the U.N. expects." Basically, Sanyal's analysis agrees with the researchers from the Spanish Foundation for Science and Technology that the trajectory of world population will most likely track the UN's low-variant trend and peak by the middle of this century.

In September 2014, demographers working with the United Nations Population Division published an article in *Science* arguing that world population stabilization is unlikely in this century. Instead, world population is projected to grow to around 11 billion by 2100. Nearly all of the projected increase—4 billion people—will happen in sub-Saharan Africa. However, the forecast basically assumes that Africa will remain an economic and political hellhole for the remainder of the century. In their November 2014 study, *World Population and Human Capital in the Twenty-First Century,* Wolfgang Lutz and his fellow demographers at the International Institute for Applied Systems Analysis counter that this prospect is unlikely. The chief difference between the two population forecasts is the issue of the education of women. The analysis done by Lutz and his colleagues takes into account the fact that the education levels of women are rising fast around the world, including in Africa. "In most societies, particularly during the process of demographic transition, women with more education have fewer children, both because they want fewer and because they find better ways to pursue their goals," they note. Given current age, sex, and educational trends, they estimate that world population will most likely peak at 9.6 billion by 2070 and begin falling. If, however, the boosting of educational levels is pursued more aggressively, then world population will instead top out at 8.9 billion in 2060 and begin dropping. Let us turn now to a fuller consideration of the theories and data that underpin the latest projections of global demographic trends.

Liberate Women, Reduce Population

So why is human fertility falling even as food has become generally more plentiful? Demographers, economists, and evolutionary psychologists have all contributed to a vast and ever-growing literature on this subject. All their explanations converge on the notion that as people become wealthier and more educated, they tend to switch from having more children to having fewer healthier and more highly educated children. Demographers call this the quantity-quality trade-off.

Falling fertility rates are overdetermined—that is, there is a plethora of mutually reinforcing data and hypotheses that explain the global downward trend. These include the effects of increased economic opportunities, more education, longer lives, greater liberty, and expanding globalization and trade, among others. The crucial point is that all of these explanations reinforce one another and synergistically accelerate the trend of falling global fertility.

Even more interestingly, all of them emphasize how the opportunities afforded women by modernity produce lower fertility. Let's briefly consider some of the fascinating contemporary research on the underlying causes of the demographic transition and what will likely happen to future human population growth.

First, recent research applying insights from evolutionary biology shows that people are not reproductive automatons driven remorselessly by blind instinct to maximize the number of their offspring, as most other species are. For example, research in 2008 by University of Michigan evolutionary biologist Bobbi Low and her colleagues analyzed the reproductive patterns of women in 170 countries. Their study, "Influences on Women's Reproductive Lives: Unexpected Ecological Underpinnings," in the journal *Cross-Cultural Research*, uses insights based on life-history theory. This approach suggests that when the risks of mortality are high, women tend to reproduce more frequently (to increase the probability of some offspring surviving to maturity) and early (to ensure reproduction before they die). In fact, Low and her colleagues found that when women can expect to live to age sixty and above, the number of children they bear falls by half. Another study in 2013 by Low and her colleagues has bolstered their finding that expecting to live beyond age sixty dramatically lowers fertility.

Looking at data from various countries confirms that fertility rates do drop as the average life expectancy of women crosses the threshold of age sixty. Consider Iran. In 1970, average life expectancy for Iranian women was fifty-four and total fertility was 6.3 children. Today, Iranian women have an average life expectancy of seventy-five and bear 1.9 children. What about Bangladesh? In 1970, female life expectancy was forty-four, and they bore 6.6 children. Today, Bangladeshi women live an average of seventy years and average 2.2 children. For India the corresponding figures for 1970 were forty-eight years and 4.9 children, and are now sixty-seven years and 2.5 children. In Brazil, female life expectancy in 1970 was sixty-one and total fertility was 4.3 children. Today, Brazilian women average seventy-eight years and total fertility stands at 1.8 children. The threshold, however, is not perfectly predictive; there are lags. Female life expectancy in Mexico was sixty-five in 1970 at a time when its total fertility rate was 5.5 children. Today, Mexican women can expect to live to about seventy-eight, and they bear 2.2 children on average. For comparison, in 1970 American women could expect to live about seventy-five years, bore 2.1 children, and infant mortality was 20 per 1,000 births. In 2012, American female life expectancy had risen to eighty-one years, births averaged 1.86 children per woman, and infant mortality had fallen to 6 per 1,000 births. By the way, the only years in which the US general fertility rate was lower than 1.86 occurred during the 1970s, when the rate fell to 1.74 births in 1976.

UN demographers expect global average life expectancy at birth to rise to seventy-six by 2050 and eighty-two by 2100. If the evolutionary biologists are right, rising life expectancy will result in falling fertility. Unfortunately, the demographers estimate that life expectancy in the world's poorest countries—many of which are severely afflicted with HIV/AIDS—is now just fifty-eight, and they project that it will reach the current global average of about seventy by 2050 and eventually rise to seventy-eight by 2100.

With regard to countries where female life expectancy is below sixty years, life-history theory frequently fails to correlate well with actual fertility rates. Some countries with low female life expectancy also have relatively low total fertility rates. For example, the increased prevalence of HIV/AIDS dramatically lowered life expectancy in a lot of African countries. According to the most recent World Bank data (2011), female life expectancy in South Africa reached sixty-two years in 1990 and has now fallen to fifty-five today. Similarly, the average Namibian woman in 1990 could expect to live to age sixty-three; that has dropped to sixty-one today. In 1990 the average life expectancy for a Zimbabwean woman was fifty-nine; it is now fifty-six. In 1990, the total fertility rates for South Africa, Namibia, and Zimbabwe were 3.7, 5.2, and 5.2, respectively. Today, despite the fact that average female life expectancy has declined, totality fertility rates in those countries have fallen to 2.4, 3.2, and 3.6 children, respectively.

On the other hand, life-history predictions with regard to fertility rates do appear to pertain to lawless countries. According to the latest World Bank data (2011), female life expectancy in Mali, Nigeria, Democratic Republic of Congo, Burundi, Ivory Coast, and Afghanistan averages at fifty-four, fifty-two, forty-nine, fifty-three, fifty, and sixty years, respectively. Their corresponding total fertility rates are 6.9, 6.0, 6.1 6.2, 4.9, and 5.4 children per woman. Social, political, and economic chaos certainly afflicts those countries. George Mason University's Center for Systemic Peace has devised a State Fragility Index as a way to measure a country's stability, with scores ranging from 0, meaning no fragility, to a high of 23, denoting a failed state.

On the index, Mali scores 19, Nigeria 16, Congo 23, Burundi 18, Ivory Coast 16, and Afghanistan 22. In contrast, all twenty-two countries with a fragility score of 0 have below replacement fertility rates. Tragically, the political violence and economic chaos endemic in so many countries in sub-Saharan Africa, the ongoing food insecurity, the pervasive risk of disease, a high before-age-five child mortality rate, the lack of education, and their low social status provide African women many grounds to wonder just how long they may expect to live. Given these uncertainties, it is little wonder that fertility rates remain high on the continent as women hedge their reproductive bets.

Further insights into how the life prospects of women shape reproductive outcomes is provided in another 2010 article in *Human Nature,* "Examining the Relationship Between Life Expectancy, Reproduction, and Educational Attainment." That study, by University of Connecticut anthropologists Nicola Bulled and Richard Sosis, confirmed Low's findings. They divvied up 193 countries into five groups by their average life expectancies. In countries where women could expect to live to between forty and fifty years, they bear an average of 5.5 children, and those with life expectancies between fifty-one and sixty-one average 4.8 children. The big drop in fertility occurs at that point. Bulled and Sosis found that when women's life expectancy rises to between sixty-one and seventy-one years, total fertility drops to 2.5 children; between seventy-one and seventy-five years, it's 2.2 children; and over seventy-five years, women average 1.7 children. The United Nations' 2012 Revision notes that global average life expectancy at birth rose from forty-seven years in 1955 to seventy years in 2010. These findings suggest that it is more than just coincidence that the average global fertility rate has fallen over that time period from 5 to 2.45 children today.

Kids Are Expensive

Research by economists further illuminates the processes that yield falling fertility. Brown University economist Oded Galor and his colleagues have devised a unified growth theory that explains how and why people begin to focus on developing and deepening their human capital (chiefly by means of education), which then further accelerates the pace of technological progress. As a result, fertility and population growth fall, enabling humanity to escape from millennia of Malthusian stagnation into the modern world of sustained economic growth.

Recall that Malthus asserted that people on average would produce as many children as they could feed, if not more. In modern econ-speak, children are a normal good: that is to say, as income increases, demand for kids would also increase.

Instead, researchers observe that as incomes increase, the number of children per woman decreases. One possible explanation for this phenomenon is that the opportunity cost of raising children has risen over time. Opportunity cost is a benefit that must be given up to acquire or achieve something else; for example, a person may have to give up a Caribbean cruise in order to be able to buy a new car. In this case, a parent would be forgoing the extra income he or she would earn working and instead spend the time rearing a child. According to this analysis, the price of children measured in forgone income rises over time, lowering demand for them.

However, Galor points out that during the initial stages of the Industrial Revolution, as incomes were increasing in Western Europe, average fertility was increasing, just as Malthus predicted. (In contrast, US fertility rates fell throughout the nineteenth century, from 7 children per white woman in 1800 to 3.5 in 1900.) Galor suggests that the opportunity cost argument for fertility decline is too simple. If income had been the key determinant, one would find that fertility should fall as any country reaches a specific level of average per capita income. Instead, Galor notes that at the end of the nineteenth century, fertility rates begin to plummet simultaneously for a number of Western European countries at very different per capita income levels.

Galor argues that fertility began to fall as Western European economies developed increased demand for human capital during what he calls the second phase of the Industrial Revolution. In this analysis, initial increases to average incomes produced by technological progress resulted in parents' increasing both the quantity and quality of their children. However, toward the end of the nineteenth century, Galor asserts, "further increases in the rate of technological progress induced a reduction in fertility, generating a decline in population growth and an increase in the average level of education."

As economic growth was increasingly fueled by the development of ever more complicated technologies and management services, the premium attached to education began to increase. The result is that parents switched from having more children to investing in fewer higher quality (more educated) children.

Galor further argues that at the turn of the twentieth century, international trade encouraged fertility rates to fall further as rich countries began to specialize in the production of the sorts of goods that required a lot of human capital to make. On the other hand, Galor contends that poor countries increasingly specialized in goods that required a lot of manual labor to produce. The result was rising income for both rich and poor countries, but a fateful divergence in fertility trends.

During the twentieth century, fertility rates basically continued to fall in rich countries as they invested in more human capital, especially in higher levels of education. In addition, as demand for human capital grew in rich countries, schooling expanded to include women, who then entered the paid workforce. This further raised the opportunity costs of having children and encouraged further reductions in fertility.

On the other hand, poor countries channeled a larger share of their gains from increased international trade into producing more children. As a consequence, "the demographic transition in these nonindustrial economies has been significantly delayed," asserts Galor, "increasing further their relative abundance of unskilled labor, enhancing their comparative disadvantage in the production of skill-intensive goods, and delaying their process of development."

OECD economist Fabrice Murtin concurs with Galor that education is the key to lower fertility rates. In his 2009 study "On the Demographic Transition," Murtin assembled data from seventy-one countries from 1870 to 2000, to conclude that "education, rather than income or health-related variables, is the most robust determinant of the birth rate, potentially explaining about 50 to 80 percent of its decrease when average schooling grows from 0 to 10 years." Galor cites data showing that the percentage of British children ages six to fourteen who were in school rose from about 10 percent in 1860 to more than 80 percent by 1895.

As noted previously, demographer Wolfgang Lutz argues that it's not just more education, but specifically more schooling for girls that correlates with deep cuts in fertility rates. For instance, the fertility rate for Ethiopian women with no formal education was 6.1 children in 2005 and 2.0 for women with secondary and higher education. Providing women access to higher education is associated with longer lives for themselves and lower child mortality. Lutz calculates that world population in 2060 would be 1 billion fewer if the education of women globally could be speeded up to the rate achieved by South Korea in the 1960s and 1970s.

As Galor noted, the demographic transition was delayed in many poor countries, but in the second half of the twentieth century these countries also began to see rapid declines in their fertility rates. Bucknell University political scientist John Doces finds that increasing international trade is now propelling the demographic transition throughout much of the developing world. In fact, as global fertility declined since the 1950s, the value of world merchandise exports during the same period has soared by nearly ninety times.

In his 2011 study "Globalization and Population: International Trade and the Demographic Transition," Doces looks at recent data from a large number of countries and finds that those that are most open to international trade are the ones experiencing the fastest decline in their fertility rates. Doces argues that the primary cost of having children is the time and money it takes to raise them, which leaves parents less time to consume other goods. International trade expands the types of goods people can enjoy and lowers their costs. The cost of rearing children does not decline substantially, so they become more expensive relative to the new opportunities and goods afforded by increased international trade.

In addition, Doces cites a 2006 study analyzing the effects on globalization on women in 180 countries that shows "increasing international exchange and communication create new opportunities for income-generating work and expose countries to norms that, in recent decades, have promoted equality for women."[46] As a result, trade-induced demand for human capital expands to include women, further cutting fertility rates in poor countries. This conclusion is further bolstered by a 2005 study by University of Helsinki economists Ulla Lehmijoki and Tapio Palokangas; according to this study, in the short run trade liberalization boosts birth rates, but in the long run it cuts fertility. Again, this is true largely because trade liberalization encourages the development of women's human capital (education), which makes childbearing relatively more costly.

The Invisible Hand of Population Control

In 2002, Seth Norton, an economics professor at Wheaton College in Illinois, published a remarkably interesting study, "Population Growth, Economic Freedom, and the Rule of Law," on the inverse relationship between prosperity and fertility. Norton compared the fertility rates of over a hundred countries with their index rankings for economic freedom and another index for the rule of law. "Fertility rate is highest for those countries that have little economic freedom and little respect for the rule of law," wrote Norton. "The relationship is a powerful one. Fertility rates are more than twice as high in countries with low levels of economic freedom and the rule of law compared to countries with high levels of those measures."

Norton found that the fertility rate in countries that ranked low on economic freedom averaged 4.27 children per woman, while countries with high economic freedom rankings had an average fertility rate of 1.82 children per woman. His results for the rule of law were similar: fertility rates in countries with low respect for the rule of law averaged 4.16, whereas countries with high respect for the rule of law had fertility rates averaging 1.55.

Economic freedom and the rule of law occur in politically and economically stable countries and produce prosperity, which dramatically increases average life expectancy and lowers child mortality; this in turn reduces the incentive to bear more children. As data from the Heritage Foundation's Index of Economic Freedom shows, average life expectancy for free countries is over eighty years, whereas it's just about sixty-three years in repressed countries.

Let's take a look at two intriguing lists. The first is a list of countries ranked on the 2013 Index of Economic Freedom issued by *The Wall Street Journal* and the Heritage Foundation. Then compare the economic freedom index rankings with a list of countries in the 2013 CIA *World Factbook* ranked by their total fertility rates. Of the thirty-five countries that are ranked as being economically free or mostly free, only two have fertility rates above 2.1—the United Arab Emirates at 2.36 and Jordan at 3.61. If one adds the next fifty countries that are ranked as moderately free, one finds that only five out of eighty-five countries have fertility rates above 3, all of them in sub-Saharan Africa except Jordan. It should be noted that low fertility rates can also be found in more repressive countries as well—for example, China at 1.55, Cuba at 1.46, Iran at 1.85, and Russia at 1.61.

In addition, along with increased prosperity comes more education for women, opening up more productive opportunities for them in the cash economy. This increases the opportunity costs for staying at home to rear children. Educating children to meet the productive challenges of growing economies also becomes more expensive and time consuming.

Thailand's experience over the past thirty years exemplifies this process. During that time, female literacy rose to 90 percent; 50 percent of the workforce is now female; and fertility fell from 6 children per woman in the 1960s to 1.5 today. Although Thailand is classified as only moderately free on the economic freedom index, its gross domestic product (GDP) grew in terms of purchasing power parity from just over $1,000 per capita in 1960 to over $8,500 per capita in 2012.

Back in 1968, Garrett Hardin declared, "There is no prosperous population in the world today that has, and has had for some time, a growth rate of zero." That's no longer true. Japan is now experiencing a fall in its population due largely to reduced fertility, as are Germany, Russia, Italy, Poland, and some 20 other countries and territories. And as we have seen, the global total fertility rate is rapidly decelerating. Of the 231 countries and territories listed in the 2013 CIA *World Factbook*, 122 are at or below replacement fertility rates.

Norton persuasively argues that Hardin's fears of a population tragedy of the commons are actually realized when the invisible hand of economic freedom is shackled. Many poor countries have weakly specified and enforced property rights. Poor property rights means that many resources are effectively left in open-access commons where the incentive is to grab what one can before another individual gets it. Norton points out that in such situations, more children mean more hands for grabbing unowned and unprotected resources such as water, fodder, timber, fish, and pastures, and for the clearing of land. Lacking the institutional incentives to invest in and preserve resources, this drive to take as much as possible as quickly as possible leads to perpetual poverty.

And what about in the past? Haven't societies collapsed due to overpopulation? To the extent it is true that some societies have suffered collapses, we now know that it was because they lacked the proper social, political, and economic institutions for channeling individual striving into a process of economic growth that ultimately promotes the accumulation of human capital and lower fertility. Very few, if any, earlier societies could be characterized as either economically free or respectful of the rule of law. Throughout history, most people lived in the institutional equivalents of open-access commons overseen by rapacious elites who encouraged high fertility rates and the plundering of natural resources. It turns out that economic freedom and the rule of law are the equivalent of enclosing the open-access breeding commons, causing parents to bear more and more of the costs of rearing children. In other words, economic freedom actually serves as an invisible hand of population control.

Hope for Africa?

The United Nations' 2012 Revision forecasts that more than half of global population growth between now and 2050 will take place in Africa, rising from 1.1 billion to 2.4 billion. The middle-variant trend for sub-Saharan Africa projects that total fertility rate will fall from 4.9 children now to 3.1 by 2050, reaching 2.1 by 2100. As noted above, a more worrying study published in an October 2014 issue of the journal *Science* suggested that by 2100 Africa's population would grow even faster, rising from 1.1 billion to between 3.1 and 5.7 billion, with a median projection of 4.2 billion.

But is it plausible that much of Africa and many of the other least developed countries will remain high-fertility basket cases for the next several decades while the rest of the world modernizes, with concomitant improvements in the life prospects of women? Surely it is reasonable to expect that new medicines, vastly more productive crops and farming techniques, high quality education delivered via low-cost computerized tablets, cheap decentralized energy, and 3-D printing of tools and goods will spill over from the labs and factories of rich countries. These modern tools will go a long way toward ameliorating the chaos and poverty currently afflicting the least developed nations. In addition, the continuing global abatement of violent conflict is already taking hold in Africa and in other poor countries. For example, in October 2014, U.S. Naval War College researcher David Burbach and Tulane University political scientist Christopher Fettweis pointed out that "after the year 2000, conflict in Africa declined, probably to the lowest levels ever." While they noted an uptick in battle deaths on the continent between 2010 and 2013, those casualties were still almost 90 percent lower than the average in the last decade of the twentieth century. "Changes in external support and intervention, and the spread of global norms regarding armed conflict, have been most decisive in reducing the levels of warfare in the continent," they concluded. "Consequently, there is no Africa exception to the systemic shift toward lower levels of armed conflict."

Among the regions of the world, according to UNESCO, adult and youth illiteracy are highest in sub-Saharan Africa. For example, among Africans aged fifteen through twenty-four, the male literacy rate stands at 76 percent versus the female rate of 64 percent. Obviously, the more that donors from rich countries can do to promote the education of women in the world's poorest countries, the better. In addition, Africa is rapidly urbanizing, which will also push fertility rates lower. For example, a 2000 study by researchers at Pennsylvania State University and Tulane University reported that African urbanites had nearly two fewer children than did their rural counterparts. Demographers working for the International Food Policy Research Institute found in 2012 an even greater urban-rural differential; Ethiopian women in the countryside had a total fertility rate of 6 offspring, whereas their city sisters averaged only 2.4 children.

This process of modernization will bring dramatic improvements in health and longer lives, resulting in a steep decline in fertility rates. Consider that the life expectancy of Bangladeshi women rose from forty-four in 1970 to seventy-two today. In addition, literacy among Bangladeshi women aged fifteen to twenty-four climbed steeply from 38 percent in 1991 to 80 percent today, actually surpassing the male rate of 77 percent. Consequently, the country's total fertility rate fell in just the twenty years between 1980 and 2000 from 6.6 to 2.9 children. It is now 2.2 children. If the current high fertility countries can realize the bare elements of political stability and economic freedom attained by Bangladesh, with its $750 per capita income, women will live longer and have fewer children. If this analysis is right—and most of the evidence points to that conclusion—the latest UN population projections will turn out to be too high and the musty Malthusian specter will finally fade away.

In his 2013 book *The Infinite Resource: The Power of Ideas on a Finite Planet,* technologist Ramez Naam asks an intriguing question: "Would your life be better off if only half as many people had lived before you?" In this thought experiment, you don't get to pick which people are never born. Perhaps there would have been no Newton,

Edison, or Pasteur, no Socrates, Shakespeare, or Jefferson. "Each additional idea is a gift to the future," Naam writes. "Each additional idea *producer* is a source of wealth for future generations." Fewer people mean fewer new ideas about how to improve humanity's lot.

Instead of disdaining fellow human beings as a cancer or a plague, as modern neo-Malthusians do, Naam rightly argues, "If we fix our economic system and invest in the human capital of the poor, then we should welcome every new person born as [a] source of betterment for our world and all of us on it."

2

Is the World Running on Empty?

IN 2010, I WAS EXCITED TO VISIT the Great Plains Synfuels Plant near Beulah, North Dakota. Why? Because I have some small bit of personal history with it. Back in 1979, I was a low-level federal natural gas regulator and a peripheral member of the team that was guiding the plant's initial development at the Department of Energy. I functioned as a petty Igor to President Jimmy Carter's energy Dr. Frankenstein. I was eager to see up close what I helped in some minor way bring to life. Why was the plant built? Because we were in the midst of an "energy crisis," since the world was running out of oil.

One proposed solution to the energy crisis was to turn America's vast reserves of coal into methane. So Congress created the Synfuels Corporation, endowing it with $20 billion, with its goal being to eventually build as many as twenty-two enormous coal gasification plants, each one producing 300 million cubic feet of natural gas per day. Since coal gasification was an unproven technology in the United States, natural gas pipeline companies were reluctant to invest in it. The federal government rushed to the rescue with massive subsidies. To make a long story short, the oil crisis passed and the $2.1 billion plant was losing money. In 1988, the Department of Energy sold it to a local utility for 4 cents on the dollar, and even now it still barely makes a profit.

Besides its relevance to my personal history, the Great Plains Synfuels Plant stands as a very apt cautionary tale about massive energy projects promoted and financed by visionary presidents and their equally visionary helpmeets in Congress and in the federal energy bureaucracies. One other thing: Just imagine how much more massive US greenhouse gas emissions would now be if all twenty-two gigantic coal gasification plants had been built. A misguided effort to solve a spurious resource depletion crisis would have had the unintended side effect of making global warming considerably worse.

In 1972, *The Limits to Growth*, a report to the Club of Rome, was released with great fanfare at a conference at the Smithsonian Institution. In a front-page article, *The New York Times* hailed the report describing the collapse of civilization as "a grim inevitability if society continues its present dedication to growth and 'progress.'" The study was based on a computer model developed by researchers at the Massachusetts Institute of Technology and designed "to investigate five major trends of global concern—accelerating industrial development, rapid population growth, widespread malnutrition, depletion of non-renewable resources, and a deteriorating environment." The goal was to use the model to explore the increasingly dire "predicament of mankind." The researchers modestly acknowledged that their model was "like every other model, imperfect, oversimplified, and unfinished."

Yet even with this caveat, the MIT researchers concluded, "If present growth trends in world population, industrialization, pollution, food production, and resource depletion continue unchanged, the limits to growth on this planet will be reached sometime within the next one hundred years." With considerable understatement, they added, "The most probable result will be a rather sudden and uncontrollable decline in both population and industrial capacity." In other words: a massive population crash in a starving, polluted, resource-depleted world.

Probably the most notorious projections from the MIT computer model involved the future of nonrenewable resources. The researchers warned: "Given present resource consumption rates and the projected increase in these rates, the great majority of currently nonrenewable resources will be extremely expensive 100 years from now." To emphasize the point, they pointed out that the prices of "those resources with the shortest static reserve indices have already begun to increase." For example, they noted that the price of mercury had increased 500 percent in the last twenty years and the price of lead was up 300 percent over the past thirty years. The advent of the "oil crises" of the 1970s lent great credibility to these projections.

The Club of Rome analysts were not alone in their fears of imminent resource depletion. In 1980, *The Global 2000 Report to the President of the United States*, issued by President Jimmy Carter's Council on Environmental Quality, basically endorsed *The Limits to Growth* projections. "If present trends continue, the world in 2000 will be more crowded, more polluted, less stable ecologically, and more vulnerable to disruption than the world we live in now. Serious stresses involving population, resources, and environment are clearly visible ahead. Despite greater material output, the world's people will be poorer in many ways than they are today," began the report.

Peak Everything?

Let's illustrate the temper of the time by recalling the legendary bet on the future prices of five commodity metals made back in 1980. That bet resonates powerfully in the ongoing fight between neo-Malthusian doomsters and cornucopian optimists. On one side, gambling that prices would spiral ever upward as a growing population used up the world's resources, stood arch-doomster Paul Ehrlich and his two acolytes, John Holdren and John Harte. Holdren is now President Obama's chief science adviser and Harte is a professor in the College of Natural Resources at the University of California at Berkeley.

On the other side of the bet stood University of Maryland doom-slaying economist and author of *The Ultimate Resource* Julian Simon. Ehrlich deeply disliked Simon, who argued that the global trend of economic data clearly showed that human ingenuity was continuously increasing the supplies and substitutes for natural resources and relent-lessly lowering their prices. So the confident claque of doomsters chal-lenged Simon to a bet on the future prices of five different metals.

In October 1980, Ehrlich and Simon drew up a futures contract obligating Simon to sell Ehrlich the same quantities that could be purchased for $1,000 of five metals (copper, chromium, nickel, tin, and tungsten) ten years later at inflation-adjusted 1980 prices. If the com-bined prices rose above $1,000, Simon would pay the difference. If they fell below $1,000, Ehrlich would pay Simon the difference. Ehrlich mailed Simon a check for $576.07 in October 1990. There was no note in the letter. The cornucopian Simon won.

As it happens, the dire predictions of the doomsters were seemingly being validated as the prices for a wide array of commodities includ-ing grain, oil, and various minerals soared in the early 1970s. But that changed almost as soon as the bet was laid. Most commodity prices started drifting downward over the next two decades.

By 1992, even the alarmist Worldwatch Institute admitted that "recent trends in price and availability suggest that for most minerals we are a long way from running out. Regular improvements in exploitative technology have allowed production of growing amounts at declining prices." A 2005 report prepared for a meeting of the developing nation members of the Group of 77 nicely summarizes what happened to com-modity prices. The report cited data from the United Nations Confer-ence on Trade and Development indicating that "over the 24 years from 1977 to 2001, real prices declined for 41 out of 46 leading commodities." The report further noted that according to the World Bank, "real com-modity prices declined significantly from 1980 to 2002, with the World

Bank's index for commodity prices down 47 percent and metal and mineral prices down 35 percent." The five commodities for which prices had increased or stayed flat were pepper, plywood, nonconiferous logs, tropical lumber, and zinc. The prices for staples such as wheat, corn, rice, cotton, wool, iron, aluminum, tungsten, tin, copper, and even crude oil were all steeply lower than they had been in the mid-1970s.

The downward drift in commodity prices did not continue. The last decade has seen sharp increases in the prices of grains, metals, and crude oil. As an April 2012 International Monetary Fund report noted, "By the end of 2011, average prices for energy and base metals in real terms were three times as high as just a decade ago, approaching or surpassing their record levels over the past four decades. Food and raw material prices also rose markedly, although they remain well below the highs reached in the 1970s." The International Monetary Fund's food price index has dropped from its 2011 high of 192 points to 147 points in February 2015, a fall of about 25 percent. Was this rapid run-up in prices a sign that the limits to growth were now upon us? Certainly many ideological environmentalists have interpreted the price spikes that way.

Peak Commodity Super-Cycle?

"The world is at, nearing, or past the points of peak production of a number of critical nonrenewable resources—including oil, natural gas, and coal, as well as many economically important minerals ranging from antimony to zinc," warned prominent environmentalist Richard Heinberg in his 2010 article "Beyond the Limits to Growth." Heinberg had earlier made plain his collapsist beliefs in his 2007 book *Peak Everything: Waking Up to a Century of Declines.* In 2012, Michael Klare, Hampshire College political scientist and defense correspondent for *The Nation,* piled on in his book *The Race for What's Left: The Global Scramble for the World's Last Resources.* "Government and corporate officials recognize that existing reserves are being depleted at a terrifying pace and will be largely exhausted in the not-too-distant future," declared Klare. In the wake of the recent hike in grain prices, long-time prophet of impending famine Lester Brown once again in 2012 declared, "The world is in transition from an era of food abundance to one of scarcity."

Don't Heinberg, Klare, Brown, and other depletionists have a point? Isn't the increase in prices a signal of emerging scarcity? Resource optimist Simon would have badly lost his bet with Ehrlich and his associates had it run between 2003 and 2013. Simon would have owed the depletionists $2,658 (in 2013 dollars). In other words, the price of the basket of five metals had increased by nearly 270 percent. So is the end really nigh this time? Most likely not.

In the 1950s, economists Raul Prebisch and Hans Singer made the seminal observation that commodity prices had been falling for many decades relative to the prices of manufactured goods. As Singer put it, "It is a matter of historical fact that ever since the [eighteen] seventies the trend of prices has been heavily against sellers of food and raw materials and in favor of the sellers of manufactured articles." The recent upsurges in commodity prices appear to contradict this trend.

"Once—maybe twice—in every generation, the global economy witnesses a protracted and widespread commodity boom. And in each boom, the common perception is that the world is quickly running out of key materials," observes David Jacks, an economist at Simon Fraser University. We are now in just such a situation. Jacks studies the phenomenon of economic "super-cycles," in which commodity prices rise and fall over periods lasting between thirty and forty years. In 2013, Jacks analyzed the price trends for thirty different commodities during the past 160 years. He finds that fifteen of the thirty commodities he tracked over the past 160 years are in the midst of super-cycles that started in the mid-1990s. In other words, the expansionary phase of the current super-cycle has run nearly twenty years so far.

Jacks also frames a useful distinction between "commodities to be grown" and "commodities in the ground." The astonishing fact is that as world population since 1850 grew sixfold and the world's economy expanded more than hundredfold, Jacks found that the prices of commodities that are grown—grains, cotton, wool, and so forth—have generally been falling. On the other hand, commodities that come out of ground—oil, tin, iron, chromium and so forth—have remained flat or have been slowly rising.

Price is determined by supply *and demand*. Between 2002 and 2007, global economic growth was the strongest and longest lasting since the 1970s. The huge boom in the prices for all sorts of resources in the current super-cycle have been chiefly generated by rising demand in fast-growing emerging economies in countries like China and India.

In their 2012 study "Super-Cycles of Commodity Prices Since the Mid-Nineteenth Century," economists Bilge Erten and José Antonio Ocampo, from Northeastern University and Columbia University, respectively, confirm that the recent price increases in commodities are the result of a super-cycle upswing. Parsing real price data for nonfuel commodities such as food and metals from 1865 to 2009, they find evidence of four past super-cycles ranging in length between thirty and forty years. The cycles they identify ran from 1894 to 1932, peaking in 1917; from 1932 to 1971, peaking in 1951; from 1971 to 1999, peaking in 1973; and the post-2000 episode, which is ongoing. The increases in commodity prices during these cycles are driven largely by increases in demand arising from strong periods of industrialization and urbanization such as those experienced by Great Britain, Germany, and the United States in the nineteenth century, Japan in the twentieth century, and China and other emerging economies at the beginning of the twenty-first century.

The super-cycles are driven by periods of accelerating economic growth that boosts demand for commodities, thus pushing up their prices. Rising commodity prices in turn encourage the development of more supplies and the invention of resource-conserving technologies. As economic growth slows down during the second part of a super-cycle, the real prices of the now copiously supplied commodities fall. In fact, the researchers find that the prices for nonoil commodities do not generally recover to their preboom averages. Before the recent fourth super-cycle upsurge, nonfuel commodity prices had fallen by a cumulative 47 percent over the past hundred years.

The Economist magazine has developed a widely cited commodities index that tracks the real prices of an extensive variety of mineral and agricultural goods. "Since 1871, the *Economist* industrial commodity-price index has sunk to roughly half its value in real terms, seeing annual average compound growth of −0.5 percent per year over the ensuing 140 years," pointed out Council on Foreign Relations energy adjunct fellow Blake Clayton in 2013. He added, "Even after the boom years of the 2000s—in 2008, for instance, as commodity indexes soared, the *Economist* index never climbed more than halfway above where it stood 163 years earlier, in real terms."

Figuring out when a super-cycle has topped or bottomed out is a fraught exercise. Nevertheless, many researchers believe that the current super-cycle in commodity prices has peaked and will soon move into its downward phase. By February 2015 the International Monetary Fund's commodity index has fallen by about 57 percent from its July 2008 peak. If the past is any guide, commodity prices could well fall to levels even lower than the price nadir of the 1990s as the expansionary phase of the current super-cycle begins to fade.

In the meantime, let's take a look at some specific depletionist predictions.

Peak Oil

Predictions of imminent catastrophic depletion are almost as old as the oil industry. An 1855 advertisement for Kier's Rock Oil, a patent medicine whose key ingredient was the petroleum bubbling up from salt wells near Pittsburgh, urged customers to buy soon before "this wonderful product is depleted from Nature's laboratory." The ad appeared four years before Pennsylvania's first oil well was drilled. In 1919, David White of the US Geological Survey (USGS) predicted that world oil production would peak in nine years. And in 1943 the Standard Oil geologist Wallace Pratt calculated that the world would ultimately produce 600 billion barrels of oil. (In fact, more than 1 trillion barrels of oil had been pumped by 2006.)

In his 1971 Sierra Club book *Energy: A Crisis in Power,* John Holdren declared that "it is fair to conclude that under almost any assumptions, the supplies of crude petroleum and natural gas are severely limited. The bulk of energy likely to flow from these sources may have been tapped within the lifetime of many of the present population." This sounds very much like the later prognostications of "peak oil" prophets.

In 1972, *The Limits to Growth* estimated known global oil reserves at 455 billion barrels. The report projected that, assuming consumption remained flat, all known oil reserves would be entirely consumed in just thirty-one years. With exponential growth in consumption, it added, all the known oil reserves would be consumed in twenty years. These dour predictions seemed plausible after the Arab oil crisis of 1973 quadrupled prices from $3 to $12 per barrel (from $16 to $63 in 2014 dollars) and when the Iranian oil crisis more than doubled oil prices from $14 per barrel in 1978 to $35 per barrel by 1981 (from $74 to $185 in 2014 dollars). In July 2008, the price of oil surged to $147 per barrel ($160 in 2014 dollars).

In response, the US federal government imposed price controls on oil and gas in the 1970s and established fuel economy standards to encourage the sale of more efficient automobiles. The sense of doom did not dissolve. In 1979 Energy Secretary James Schlesinger proclaimed, "The energy future is bleak and is likely to grow bleaker in the decade ahead." *The Global 2000* warned, "By 2000 nearly 1,000 billion barrels of the world's total original petroleum resource of approximately 2,000 billion barrels will have been consumed." The report predicted that the price of oil would rise by 50 percent, reaching $100 per barrel by

2000. In fact, by 2000, the average price of oil in real dollars had fallen by two-thirds of its price in 1980 (at its low point in 1998, the price of petroleum in real terms was under a fifth of its 1980 price).

Historically, on the basis of annual average real prices, 1980 and 1981 were the two years with the highest oil prices, at $106 and $92 per barrel (in 2014 dollars). As the next two decades saw the price of crude bumpily descend from its 1980 high, fears of imminent depletion of oil provoking a permanent economic crisis abated. That changed when the price of petroleum began its dramatic rise in the early 2000s. As a consequence, 2013 now ranks third, with a real price averaging just under $92 per barrel. Not too surprisingly, the rapid ascent in the price of oil again excited predictions of its imminent depletion and ensuing disaster.

For example, in 2006 Princeton geologist Ken Deffeyes was warning that the imminent peak of global oil production would result in "war, famine, pestilence and death." Deffeyes, author of 2001's *Hubbert's Peak: The Impending World Oil Shortage* and 2005's *Beyond Oil: The View from Hubbert's Peak*, had already predicted that the peak of global oil production would occur on Thanksgiving 2005.

Deffeyes was far from alone. Houston investment banker Matthew Simmons stated in his 2005 book *Twilight in the Desert: The Coming Saudi Oil Shock and the World Economy* that the Saudi Arabians were lying about the size of their petroleum reserves, claiming that they are really running on empty. In September 2005 Simmons announced that "we could be looking at $10-a-gallon gas this winter." The price of gasoline in December 2005 was about $2.25 per gallon.

In a 2005 bet consciously modeled on the Simon-Ehrlich bet, *New York Times* columnist John Tierney and peak oil proponent Matthew Simmons wagered $5,000 on whether the price of oil in 2010 would average above $200 per barrel. When the bet was made, the price was $65 per barrel. When the bet was settled on January 1, 2011, the price of oil had increased to a 2010 average of $71 per barrel.

Colin Campbell, a former petroleum geologist who founded the Association for the Study of Peak Oil and Gas, had warned way back in

2002 that we were headed for peak oil production, and that this would lead to "war, starvation, economic recession, possibly even the extinction of homo sapiens." In his 2004 book *Out of Gas: The End of the Age of Oil*, the Caltech physicist David Goodstein wrote that the peak of world production was imminent and that "we can, all too easily, envision a dying civilization, the landscape littered with the rusting hulks of SUVs." Jim Motavalli, then editor of the environmentalist magazine *E*, wrote in the January/February 2006 issue: "It is impossible to escape the conclusion that we're steaming full speed ahead into a train wreck of monumental proportions."

And James Schlesinger, the country's first secretary of energy, declared in the Winter 2005–06 issue of the neoconservative foreign policy journal *The National Interest* that "a growing consensus accepts that the peak is not that far off." He added, "The inability readily to expand the supply of oil, given rising demand, will in the future impose a severe economic shock." A 2007 report by the German think tank Energy Watch Group (EWG) concluded that the world had reached the peak of oil production in 2006 and that supplies would fall from about 81 million barrels per day to just 39 million by 2030. "The world is at the beginning of a structural change of its economic system. This change will be triggered by declining fossil fuel supplies and will influence almost all aspects of our daily life," declared EWG founder Jörg Schindler. This fast onset of oil supply shortfalls, warns the EWG report, could trigger the "meltdown of society." As the price of petroleum ascended in July 2008 to $147 per barrel, an analysis released by the investment firm Goldman Sachs suggested that oil prices might soar to $200 per barrel. That did not happen. In fact, by the end of 2008, the price of oil had fallen to $34 per barrel.

Most of the petro-doomsters base their forecasts on the work of the geologist M. King Hubbert, who correctly predicted in 1956 that US domestic oil production in the lower forty-eight states would peak around 1970. In fact, US production did reach 9.6 million barrels per day in 1970 and then began declining. In 1969, Hubbert predicted that world oil production would peak around 2000.

Hubbert's Peak

Hubbert argued that oil production grows until half the recoverable resources in a field have been extracted, after which production falls off at essentially the same rate at which it expanded. This theory suggests a bell-shaped curve rising from first discovery to peak and descending to depletion. Hubbert calculated that peak oil production follows peak oil discovery with a time lag. Globally, discoveries of new oil fields peaked in 1962. The time lag between peak global discoveries and peak production was estimated to be around thirty-two years, but peak oilers claim that the two oil crises of the 1970s reduced consumption and thereby delayed the peak until now. Hubbert's modern disciples argue that humanity has now used up half of the world's ultimately recoverable reserves of oil, which means we are at or over the peak.

Have we reached or passed the halfway mark of world petroleum reserves? Not yet. The *2014 Statistical Review of World Energy*, issued by the oil company BP, notes that global proved oil reserves rose from 1.04 billion barrels in 1992 to 1.69 trillion barrels in 2011. The US Energy Information Administration's *2013 International Energy Outlook* concurs with the BP review and reports that known world oil reserves total 1.6 trillion barrels.

Recall that peak oil was supposed to have happened in 2005 (on Thanksgiving Day!), when production was 82 million barrels per day. In 2014, global oil production reached 92 million barrels per day. At that rate of production, current reserves would last fifty years. In general, most experts project that to meet demand in 2035, world oil production will rise to around 110 million barrels per day. At that rate, known reserves would last forty years. In addition, proved reserves of natural gas in 2012 stood at 187 trillion cubic meters and annual production was 3.4 trillion cubic meters, at which rate known supplies would last fifty-five years.

The International Energy Agency annually issues its *Resources to Reserves* report. The IEA is an international think tank established in the 1970s by the governments of the rich countries that are members of the Organisation for Economic Co-operation and Development. The IEA advises those governments on energy supplies and policies. Like most organizations that evaluate future resource supplies, the IEA makes a distinction between reserves and resources. Reserves are known stocks that can be produced economically using today's technology. Resources are stock estimates that are judged likely to be ultimately producible depending on market and technological developments.

The IEA's *Resources to Reserves 2013* report estimated that worldwide proven conventional oil reserves stand at 1.3 trillion barrels. Adding estimated recoverable conventional oil resources brings the total to 2.7 trillion barrels. The IEA then estimates that reserves of unconventional oil are around 400 billion barrels and recoverable resources of 3.2 trillion barrels. This assessment brings the total world petroleum reserves to 1.7 trillion barrels and the global resource base to just under 6 trillion barrels. In his 2012 report *Oil: The Next Revolution*, Harvard University Belfer Center scholar Leonardo Maugeri agrees: "Oil is not in short supply. From a purely physical point of view, there are huge volumes of conventional and unconventional oils still to be developed, with no 'peak-oil' in sight." As evidence, he cites data from the US Geological Survey suggesting that the remaining conventional oil resources are 7 to 8 trillion barrels.

With regard to natural gas supplies, thanks in large measure to the combination of hydro-fracturing and horizontal drilling, vast new stores have been released from previously unexploitable shale formations. The IEA in 2011 issued a report asserting: "Conventional recoverable resources are equivalent to more than 120 years of current global consumption, while total recoverable resources could sustain today's production for over 250 years." As the IEA 2013 report succinctly notes, "Fossil fuels are abundant in many regions of the world and they are in sufficient quantities to meet expected increasing demands."

The fundamental error made by the peak oil disciples of Hubbert is now clear; they substantially underestimated the actual amount of petroleum reserves and resources and the oil industry's ever-increasing technological prowess to exploit them. The notion that once half of the oil in a field has been produced, the only direction is down seems intuitively obvious. And oil is, after all, an exhaustible resource of which there is only so much. Peak oil theorists ominously point out that since the 1980s the volume of new discoveries has been smaller than the amount of oil extracted. Yet, oil reserves have continued to grow. How could this be?

Swedish economist Marian Radetzki explained this paradox by noting that "the quantity of reserves in new discoveries regularly appreciates in the process of field development exploration and subsequent exploitation." In other words, oil companies often find more crude in their wells than initially predicted. In addition, peak oilists tend to assume that oil production technology is static when in fact constant improvement enables the extraction of ever-greater quantities of oil from a field. "Historical data from the United States reveal that the ultimately recovered oil when a field ceases to produce is on average six times as large as the volume announced after the initial discovery," observes Radetzki. Oil companies basically look for more crude in fields that that they have already discovered and regularly find it in the form of appreciating reserves.

For example, since 1950 some 2.6 million oil and gas wells have been drilled in the United States and more than 800,000 are currently producing oil or gas. In contrast, Saudi Arabia has only about 2,900 operating wells, and all of the Organization of Petroleum Exporting Countries (OPEC) wells total just 37,500. This suggests, even taking into account differences in geology, that there is plenty of scope for boosting the production of known fields by means of reserve appreciation in OPEC and other oil-producing countries.

Consider the Kern River Field in California, which was discovered in 1899. In 1942 it was estimated that only 54 million barrels remained to be produced there. During the next forty-four years the field produced 736 million barrels and had another 970 million barrels remaining. In 1980 the US was estimated to have between 27 and 30 billion barrels of reserves. In 2013 the US Energy Information Administration estimated US proven oil reserves at 29 billion barrels, even though American oil fields had produced about 80 billion barrels of oil between 1980 and 2013.

Back in 1973, US Foreign Service officer James Akins dryly observed: "Oil experts, economists, and government officials who have attempted in recent years to predict the future demand and prices of oil have had only marginally better success than those who foretell the advent of earthquakes or the second coming of the Messiah." Keeping Akins's admonition firmly in mind, what does future energy demand look like?

In their April 2013 study "The Global Energy Outlook," Duke University researchers Richard Newell and Stuart Iler comprehensively review petroleum consumption and production estimates from the US Energy Information Administration, the International Energy Agency, and the oil companies ExxonMobil and BP. Those estimates more or less converge on an increase in consumption from the current 88 million barrels per day to around 110 million barrels per day in 2035. Similarly, they project that natural gas demand will rise from 320 billion cubic feet per day now to somewhere between 462 to 514 billion cubic feet per day in 2035. They do further note: "While energy consumption continues to grow, it is growing at a slower rate as energy continues to decouple from economic growth, due to structural transformation in the economy and technological improvements in energy efficiency. Fossil fuels will continue to dominate the energy mix, but their share is falling, and for the first time the absolute level of some fossil fuels looks ready to plateau and then potentially decline."

All of the energy consumption estimates reported by Newell and Iler incorporate assessments concerning likely improvements in energy efficiency and production technologies, as well as the adoption of subsidies for renewables and prices on carbon dioxide emissions. For example, ExxonMobil's *2014 Outlook for Energy* report assumes that rich developed countries will impose carbon dioxide emissions control measures that amount to "an implied cost of CO_2 emissions that will reach about $80 per tonne in 2040." Similarly, the consultancy Synapse Energy Economics projected in its 2012 report that carbon dioxide prices would range between $35 and $90 per ton by 2040.

What will the price of oil be in the future? One can find just about any estimate one wants. For example, in late 2013, Reuters polled twenty leading oil industry experts and obtained estimates for 2020 prices ranging from $70 to $160 per barrel. The International Energy Agency projects that the price of oil will be around $128 per barrel (2012 dollars) in 2035.

In their analysis of commodity super-cycles, Erten and Ocampo report: "In contrast to these trends in non-oil commodity prices, real oil prices have experienced a long-term upward trend, which was only interrupted temporarily during some four decades of the twentieth century." This suggests that the price of oil will not likely fall back to its 1998 average of $17 per barrel (2014 dollars). High oil prices have, however, drawn forth substantial investments in new production. Later in this decade, extra supplies of crude will meet weakening demand as the super-cycle decelerates, resulting in falling prices.

Harvard University analyst Maugeri also argues that recent big investments in production and innovation will enable oil producers to bring an additional 18 million barrels of oil per day to the market by 2020, raising global production to around 110 million barrels per day. "The age of 'cheap oil' is probably behind us," writes Maugeri, "but it is still uncertain what the future level of oil prices might be. Technology may turn today's expensive oil into tomorrow's cheap oil." By cheap oil, Maugeri means the $20 to $30 per barrel price that prevailed during most of the last half of the twentieth century. He believes that the trend toward more production will tend to stabilize the price of oil after 2015. Maugeri's suggestion that oil prices could dip below $50 per barrel in the near term was realized in January 2015 when the price for benchmark West Texas Intermediate crude hovered around $45 per barrel, but he generally assumes that the price will remain above $70 in the run-up to 2020.

Despite reassuring petroleum reserve estimates and the downward pressure on prices that increased production and the waning of the current commodity super-cycle generates, a peak oil crisis might still happen. How? Through political mismanagement.

Political Peak Oil

"The real problems concerning future oil production are above the surface, not beneath it, and relate to political decisions and geopolitical instability," notes Maugeri. It is a disquieting fact that government-owned oil companies control nearly 90 percent of the world's oil reserves and produce about 75 percent of current supplies. Not guided by the profit motive to take future income into account, governments use their state-owned oil companies to plunder petroleum reserves as quickly and as messily as possible, often using oil revenues to buy off restive populations. In the wake of the Arab Spring uprisings in 2011, Saudi Arabia's monarchy bought social peace by quickly boosting wages by $130 billion; Algeria announced a $156 billion infrastructure and jobs program; and Kuwait gave every citizen $3,600 and fourteen months of free groceries. In addition, tens of billions of dollars generated by government-owned oil companies are diverted into the private offshore accounts of corrupt politicians. Some scholars have dubbed this depressing dynamic the "oil curse."

Consequently, government oil companies typically underinvest, with the result that oil production is far less than is technically and economically possible. According to Maugeri, private oil companies currently recover about 35 percent of the petroleum in a typical oil field. If that recovery factor could be increased by another five percentage points, it would boost worldwide recoverable reserves by more than all of Saudi Arabia's current proven reserves. The oil recovery rates for government-owned oil companies are, however, much lower than those of private companies. "The oil recovery rate is well below 25 percent, because of old technologies, reservoir mismanagement, limited investment," explains Maugeri. He estimates that the recovery rate for several major oil-producing countries—including Russia, Iran, Venezuela, Kuwait, and Iraq—is actually below 20 percent.

As noted earlier, lots of the world's oil is unfortunately produced by government-owned companies run by corrupt regimes. Besides those just listed, there are the basket cases of Nigeria, Chad, Sudan, Angola, and Libya—and that's just in Africa. If an "oil crisis" fails to materialize, it will be chiefly because nimble private oil companies will have succeeded in boosting production capacity in enough places around the world that temporarily losing one or two major producers to incompetence or malice won't matter much. But the sad fact is that the world's energy security would be a lot greater if more of the world's oil and gas resources were in the hands of private companies. Peak oil, if it occurs, will be the result of human folly and government greed, not because the world has suddenly run out of crude.

Finally, it should be noted that many environmentalists aren't scared that we will soon run out of oil; instead, they fear that we won't. Why? Because they worry about the effect the carbon dioxide emitted into atmosphere by burning all that oil will have on the climate. Let's set that concern aside until the chapter asking "Can We Cope with the Heat?"

"Nature makes a drought, but man makes a shortage"

That's the trenchant slogan that Leiden University College water resource economist David Zetland uses to sum up how bureaucratic mismanagement of supply and demand in most countries pervasively misallocates water. He also makes the intellectually elegant point that people must not confuse scarcity and shortages. Scarcity arises from the fact that human wants are boundless while the resources to satisfy them are limited. Since we can't have it all, we manage the scarcities engendered by our desires by allocating our time and money between competing goods and services.

For example, if a person chooses to redo her kitchen, then she can't afford to jet off to a ski vacation. Demand and supply ensure that there is no shortage of cabinets or hotel rooms in Aspen. A shortage occurs when cabinets or hotel rooms are not available at any price. This generally occurs when governments push the price of a good or service below the cost of supplying them to consumers. Older Americans may remember how shortages of all kinds broke out in the 1970s when President Richard Nixon imposed wage and price controls. "Scarcity leads to shortage when water managers fail to balance supply and demand," notes Zetland.

"Scarcity and shortage are the same for water as they are for other goods—except that most other goods are traded in markets in which rising and falling prices balance supply and demand to prevent shortages," explains Zetland. Nearly every good is scarce, but in market economies, shortages for most goods rarely occur. The chief problem with water is that it is mostly supplied by government agencies or government-sanctioned monopolies whose prices are purposely held below the actual costs of supplying water. This is essentially a system of government price controls and the predictable result is a shortage. "Underpricing (or zero pricing in some cases) has sustained overuse: if markets delivered Porsche cars at give-away prices, they too would be in short supply," observes the United Nations Development Program's 2006 *Summary Human Development Report 2006—Beyond Scarcity: Power, Poverty and the Global Water Crisis.*

Zetland also notes, "The supply of water on the planet is fixed, but useful supplies are not." Unfortunately, the useful supplies are being badly misallocated. The good news is that between 1990 and 2010, over 2 billion people gained access to improved drinking water sources, such as piped supplies and protected wells. Nevertheless, some 780 million still do not have access to safe water, and as a consequence, 3,000 children die every day of preventable diarrheal diseases. Even now, 3.6 billion people—about half the world's population—still do not have access to water piped to their homes, compounds, or yards.

Claims that the world is running out of water are wrong, but it is true that foolish policies are creating and exacerbating shortages. For example, irrigation for agriculture is the biggest use for water in most of the world. Unfortunately, nearly every country subsidizes water for their farmers, thus encouraging them to overuse the resource. In the United States, resource economist Delworth Gardner, a professor emeritus at Brigham Young University, once calculated that the total cost to society of a typical federal irrigation project is $400 per acre-foot of water (an acre-foot is the amount of water it takes to cover one acre to a depth of one foot). The market value of the water ranges from $50 to $100 per acre-foot, but farmers usually pay the Bureau of Reclamation about $20 to $30. In other words, US farmers are often paying less than a tenth of the cost of supplying them with water. If water prices reflected the actual costs of supplying it, then farmers would be encouraged to conserve water by shifting to less thirsty crops or from furrow irrigation to drip irrigation. So the first step toward addressing water shortages is for governments to stop subsidizing people to waste it.

Although unraveling the current mess will be politically difficult, clearly the next step would be to allocate secure property rights in water to people who can then sell and trade water in markets much like any other good. Let's say that water supplies are allocated initially to farmers. If cities and businesses had a higher demand for water, they could then bargain with the farmer owners for secure supplies and both sides would benefit. Even better from the point of view of conservationists, water could be purchased from farmers so that it could be left in streams to protect and nurture fish and other riparian wildlife. In this way, markets make it easier to allocate water to its highest best use and for farmers to earn new income.

Can markets adequately provide safe drinking water, even to poor people? First, the plain fact is that governments and government monopolies, especially in developing countries, have been terrible at supplying water to their citizens.

In his 2006 monograph *Water for Sale: How Business and the Market Can Resolve the World's Water Crisis,* economic analyst Fredrik Segerfeldt from the Swedish think tank Timbro makes the case that water privatization can go a long way toward quenching the thirst of the poor. Segerfeldt points out that public water systems in developing countries generally supply politically connected wealthy and middle-class people, whereas the poor are not hooked up to municipal water mains. Segerfeldt cites one study of fifteen countries that found that in the poorest quarters of their populations, 80 percent of the people were not hooked up to water mains. Of course, the poor don't just die of thirst; they just pay more—generally a lot more—for their water.

"Contractors often drive tankers to poor districts, selling water by the can, in which case the very poorest of the world's inhabitants are already exposed to market forces but on very unfair terms, because water obtained like this is on average twelve times more expensive than water from regular water mains, and often still more expensive than that," notes Segerfeldt. A survey of major cities in developing countries found that the poor in Lagos, Nigeria, pay 4 to 10 times more for their water than people who are hooked up to water mains do; in Karachi, Pakistan, they pay 28 to 83 times more; in Jakarta, Indonesia, 4 to 60 times; and in Lima, Peru, 17 times more. Essentially, the rich get cheap tap water while the poor pay the moral equivalent of Perrier prices.

Some countries have now turned to the private sector and multinational companies for help in providing their thirsty poor citizens with water. Privatization can mean selling entire water supply and treatment systems to private owners; long-term leases of water supply systems; or contracts to manage public water systems. In practical terms, the usual arrangement is a long-term lease. So far, only 3 percent of the poor in developing countries get their water from private-sector water systems. However, these initial projects have provoked an outcry by anti-privatization activists around the world against a "global water grab" by giant corporations.

Segerfeldt shows that even imperfect privatization efforts have already successfully connected millions of poor people to relatively inexpensive water where government-funded efforts have failed. For example, before privatization in 1989, only 20 percent of urban dwellers in the African nation of Guinea had access to safe drinking water; by 2001, 70 percent did. The price of piped water increased from 15 cents per cubic meter to almost $1, but as Segerfeldt correctly notes, "before privatization the majority of Guineans had no access to mains water at all. They do now. And for these people, the cost of water has fallen drastically. The moral issue, then, is whether it was worth raising the price for the minority of people already connected before privatization in order to reach the 70 percent connected today." In Cartagena, Colombia, privatization boosted the number of people receiving piped water by 27 percent. Even the controversial privatization in Buenos Aires saw the number of households connected to piped water rise by 3 million; 85 percent of the new customers lived in the poor suburbs of the city. Segerfeldt cites other successful privatizations in Gabon, Cambodia, Indonesia, and Morocco.

Oddly, the freshwater chapter in the IPCC 2014 *Adaptation* report fails to mention markets or prices as a way to manage water scarcity. The report does, however, note: "Barriers to adaptation in the freshwater sector include lack of human and institutional capacity, lack of financial resources, lack of awareness, and lack of communication." Markets and prices are precisely the institutions that increase the capacity to manage scarce resources, provide financing, increase awareness, and enhance communication. It is sadly the case that there is a general tendency on the part of governments and activists to fiercely resist markets and property rights as ways to protect and control resources and environmental amenities until public mismanagement finally provokes a crisis. But as we've seen, such crises have been and will continue to be resolved by adopting the institutions of the market.

Peak Everything Redux

Back in 1972, the computer modelers for *The Limits to Growth* calculated 42 years ago that known world copper reserves would be entirely depleted in 36 years, lead in 26 years, mercury in 13 years, natural gas in 38 years, petroleum in 31 years, silver in 16 years, tin in 17 years, tungsten in 40 years, and zinc in 23 years. In other words, most of these nonrenewable resources would be entirely used up before the end of the twentieth century.

Being sensible folks and desiring to be conservative in their predictions, they recognized that it was very likely that undiscovered reserves would be found and that technological improvements in extracting resources would occur. So just to be generous, they made the same depletion calculations with known reserves increased fivefold. At exponential consumption rates that they expected to unfold, they calculated after a gratuitous fivefold increase in resources there would *now* be only 15 years of aluminum left, 8 years of copper, 1 year of mercury, 9 years of natural gas, 10 years of petroleum, 2 years of silver, 21 years of tin, and 10 years of zinc.

As noted earlier, the recent general increase in commodity prices spurred on by rising demand during the current economic super-cycle has called forth screeds warning of "the race for what's left" and "peak everything." Are the depletionists right this time? Not really.

Mineral reserves are generally defined as ores that now are economically and technically practical to extract, while resources are ores for which reasonable prospects exist for eventual economic extraction. Based on current consumption rates, the US Geological Survey in its *2014 Mineral Commodity Summaries* report estimates that the world has 108 years of reserves of bauxite, which is used to produce aluminum. The USGS further estimates that total bauxite resources amount to 75 billion tons, which would last as much as 290 years at current rates of production. Similarly, at current consumption rates, known copper reserves will last 41 years. The USGS estimates known resources of copper at about 1.8 billion tons and total resources, including undiscovered deposits, at 3.1 billion tons. This suggests that at current rates of production there is a 105-year to 182-year supply of copper. Known lead reserves will last 16 years, although the USGS estimates that lead resources equal 2 billion tons and that would mean a supply lasting somewhat more than 370 years.

Mercury reserves are enough to last another 53 years, but the USGS notes, "The declining consumption of mercury, except for small-scale gold mining, indicates that these resources are sufficient for another century or more of use." Current silver, tin, tungsten, and zinc reserves will respectively last 20, 20, 49, and 19 years more. While not making a formal estimate, the USGS notes that world tin resources "if developed, could sustain recent annual production rates well into the future." Zinc resources would last 141 years.

During the last decade the price of steel has escalated. Does this signal imminent depletion? The USGS reports that known world iron ore reserves are 170 billion tons and world resources amount to 800 billion tons. In 2013, the world mined about 3 billion tons of ore. This implies that at current rates of production known iron ore reserves and resources would last 57 years and 667 years, respectively. Other metals are used to harden iron into steel, including chromium and manganese, both of which have also experienced an increase in their prices. With regard to chromium, the USGS simply notes: "World resources are greater than 12 billion tons of shipping-grade chromite [chromium ore], sufficient to meet conceivable demand for centuries." Known reserves of manganese amount to 570 million tons, of which 17 million were mined in 2013. This suggests that known reserves could supply world demand for 33 years. The USGS just states that "land-based manganese resources are large."

Why does the horizon of known mineral reserves never go out further than a few decades? Basically because miners and technologists do not find it worthwhile to discover new sources and develop new production techniques until markets signal that they are needed. How this process evolves is encapsulated by the 2014 USGS report, which notes: "In 1970, identified and undiscovered world copper resources were estimated to contain 1.6 billion metric tons of copper, with reserves of about 280 million metric tons of copper. Since then, almost 460 million metric tons of copper have been produced worldwide, but world copper reserves in 2013 were estimated to be 690 million metric tons of copper, more than double those in 1970, despite the depletion by mining of more than the original estimated reserves."

Ever More Peaks

Based on USGS data, stocks of industrially important metals do not seem to be in imminent short supply, but are we on the verge of running out of other critical nonrenewable resources? General commodity price increases generated by the current economic super-cycle has certainly drawn a lot of would-be diviners of disastrous depletion into the public square.

Consider recent alarms about "peak fertilizer." In the 1840s German chemist Justus von Liebig became the "father of the fertilizer industry" when he discovered that applying nitrogen, phosphorus, and potash to crops boosted yields. Soon farmers began spreading phosphorus-rich imported guano, obtained from the ancient congealed droppings of seabirds found on deserted oceanic islands, on their fields. Now phosphorus and potash (potassium oxide) are mined.

In their April 20, 2010, article "Peak Phosphorus," in the journal *Foreign Policy*, James Elser and Stuart White warned that there may be only enough phosphorus to satisfy agricultural demands for the next thirty to forty years. Failing to meet the challenge of peak phosphorus would mean, they argued, that "humanity faces a Malthusian trap of widespread famine on a scale that we have not yet experienced." In 2012, famed financier Jeremy Grantham joined the peak phosphorus parade and added peak potash to the cavalcade in a column published in the science journal *Nature*. "There is the impending shortage of two fertilizers: phosphorus (phosphate) and potassium (potash). These two elements cannot be made, cannot be substituted, are necessary to grow all life forms, and are mined and depleted. It's a scary set of statements," warned Grantham. "What happens when these fertilizers run out is a question I can't get satisfactorily answered and, believe me, I have tried."

Evidently, Grantham and others alarmed about the possibility of running out of fertilizer did not look very hard for satisfactory answers to worries about peak phosphorus and potash. The USGS estimates that current reserves of phosphate rock will last 300 years at current rates of production. The estimated total resources of phosphate rock would last over 1,300 years. Similarly, known supplies of potash would last 101 years at current rates of usage. Total world estimates of potash resources would last 7,000 years. And finding and applying more phosphorus is not the only way to boost yields.

Biotechnologists are exploring ways to design crop plants that greatly increase the efficiency with which they use phosphorus. In recent low-phosphorus experiments biotech-enhanced soybeans produced nearly fifty more seeds than conventional varieties. The development of such nutrient-efficient crops would substantially reduce the volume of fertilizer needed to grow a given amount of food, and ease whatever pressures there might be on reserves of the minerals. The invention of resource-sparing technologies is a pervasive feature of the modern economy, empowering people to get ever more value out of less material.

Other peak prognosticators focused their attention on lithium. Lithium is the element at the heart of the batteries that green-energy enthusiasts hope will spark an electric car revolution. But in 2007, William Tahil, a researcher with the France-based consultancy Meridian International Research, fueled the "peak lithium" meme with his report, "The Trouble with Lithium." Tahil's analysis alarmingly concluded that there is "insufficient economically recoverable lithium available in the Earth's crust to sustain electric vehicle manufacture in the volumes required." Tahil added, "Depletion rates would exceed current oil depletion rates and switch dependency from one diminishing resource to another." Is Tahil right?

Looking once again at USGS reserve and resource assessments, one finds that the agency estimates that at current production rates the reserves of lithium would last 371 years and the estimated resources would last 1,142 years. Of course, if demand for electric vehicles takes off, production would have to increase significantly. Perhaps a good way to think about future demand for lithium is to consider that 9 kilograms (about 20 pounds) of it goes into the Tesla Model S 990-pound battery pack. The Tesla can go about 250 miles on a single electric charge. A ton of lithium is enough to produce 111 Tesla batteries, and in 2013 the world produced 35,000 metric tons of lithium. Current lithium production could therefore notionally supply batteries to power just under 3.9 million Tesla Model S cars. In 2013, US-based automakers produced just over 11 million vehicles. Assuming that all used the same batteries as the Tesla Model S implies a consumption of about 100,000 metric tons of lithium annually. At that rate, reserves would last 123 years, and estimated resources 380 years.

A 2011 study on global lithium availability by researchers at the University of Michigan and Ford Motor Company estimated that the cumulative twenty-first-century demand for lithium would likely range between 12 and 20 million tons, depending on assumptions regarding economic growth and recycling rates. "Even with a rapid and widespread adoption of electric vehicles powered by lithium-ion batteries, lithium resources are sufficient to support demand until at least the end of this century," concluded the researchers. Similarly, a 2011 study of future lithium demand and supply by researchers at the University of California at Berkeley concluded: "Eventually, on the order of 1 billion 40 kWh Li-based EV batteries can be built with the currently estimated reserve base of lithium."

In this supposed "century of declines," it is child's play for peddlers of the limits-to-growth meme to seize on any hike in the price of some raw material and solemnly declare that it signals "peak whatever." And so it was when the prices of rare earth metals like neodymium and dysprosium began to ascend. Let's look specifically at neodymium. It is used extensively to produce permanent magnets found in everything from magnetic disk readers and cell phones to wind turbines and automobiles. For example, the magnets that drive a Prius hybrid's electric motor use more than two pounds of neodymium. Interestingly, neodymium magnets were invented in the 1980s to overcome the global cobalt supply shock that occurred as the result of internal warfare in Zaire. Because China can more cheaply produce neodymium than any other country in the world, that country is the source of about 90 percent of the world's neodymium. In 2010, China's government warned that it would begin restricting exports of neodymium (and other rare earth metals) in order to ensure supplies for its own manufacturers.

The announcement of China's intended export restrictions unsurprisingly excited peakists into spreading alarm about peak rare earths. To counteract the supposedly impending global shortages, Rep. Mike Coffman (R-CO) introduced in March 2010 his Rare Earths Supply-Chain Technology and Resources Transformation (RESTART) Act. The RESTART Act would offer federal loan guarantees to mining and refining companies to re-create in five years a domestic rare earth minerals industry. Rare earth minerals independence, if you will. The bill apparently died from lack of action in 2011, perhaps because the prices of rare metals fell fast. "Global market forces are leading to positive changes in rare earth supply chains and a sufficient supply of most of these materials likely will be available to the defense industrial base," noted a comprehensive review of industrial raw materials supplies by the Department of Defense in late 2013. The report further observed, "Prices for most rare earth oxides and metals have declined approximately 60 percent from their peaks in the summer of 2011." Why did prices fall? As the DOD review reports, "One factor contributing to reduced demand is the substitution of other materials for rare earth materials." For example, Tesla Motors installs induction motors that do not use rare earths; LumiSands in Washington State has developed LED lights that replace rare earths with abundant silicon.

In addition, China's threat to create a rare earths cartel provoked exploration and the opening of new mines around the world, including in the United States, Australia, and Malaysia. "There are over 400 rare earth projects under review globally, approximately four dozen of which may be considered in advanced stages of development in over a dozen countries worldwide," notes the DOD review. This is exactly the response that one would expect to higher prices. By January 2015, rare earth prices were 8 percent below their 2011 highs.

Kudos must be given to tireless and imaginative peakists for their astonishing ability to foment alarm about the future availability of rare earth metals. The USGS estimates that at current levels of consumption, the known reserves of rare earths will last 1,272 years. The agency doesn't bother with providing figures for the ultimately recoverable resource base, simply noting that "undiscovered resources are thought to be very large relative to expected demand."

Proponents of peak depletion get it wrong because they treat natural resources as fixed stocks, failing to take into account the inherent dynamics of market forces and technological innovation. Amazingly, some still claim that the era of cheap resources is over, when in point of fact nearly all resources in the past were much more expensive than they are today, even taking into account the current super-cycle.

Resources are defined by advancing human knowledge and technology. A deposit of copper is just a bunch of rocks without the know-how to mine, mill, refine, shape, ship, and market it. "Innovation has arguably been the dominant force in determining the path of real prices for primary commodities over the past three and a half centuries," assert economists Harry Bloch and David Sapsford. They add, "The influence of innovation has been sufficient to result in negative trends in real prices for numerous individual commodities and for aggregate indexes of commodities. The negative trends occurred in spite of massive increases in output with growth in the world economy."

Richer Means Cleaner

The folks who put together *The Limits to Growth* back in 1972 concluded that if humans were somehow able to overcome all other "limits," pollution would still do us all in. "Virtually every pollutant that has been measured as a function of time appears to be increasing exponentially," read the report. It turns out that they were making this exponentialist prediction just as a wealthier United States was reaching the per capita income thresholds at which citizens begin to demand better environmental quality. Happily, once again, the new Malthusians had things exactly backward.

Consider that the EPA reports that between 1980 and 2011, US gross domestic product increased 128 percent, vehicle miles traveled increased 94 percent, energy consumption increased 26 percent, and the US population grew by 37 percent. During the same time period, total emissions of the six principal air pollutants dropped by 63 percent.

Why does pollution peak and then fall? More than two decades ago, economics scholars noted that when incomes begin to rise, pollution gets worse—until it doesn't. Income and pollution data from around the world have revealed that there are various per capita income thresholds at which air and water pollutants begin to decline. This discovery has been dubbed the Environmental Kuznets Curve (EKC). The Environmental Kuznets Curve hypothesis posits that environmental conditions initially deteriorate as economic growth takes off, but later improve when citizens with rising incomes demand better quality environmental amenities. There is still considerable debate over the empirical reality of this hypothesis, but a 2011 meta-analysis based on 878 observations from 103 empirical EKC studies (1992 to 2009) reports that its results "indicate the presence of an EKC-type relationship for landscape degradation, water pollution, agricultural wastes, municipal-related wastes, and several air pollution measures." The best evidence backs the notion that increasing wealth from economic growth correlates with a cleaner natural environment—that is to say, richer becomes cleaner.

While levels for many pollutants are falling in rich developed countries, it must be acknowledged that globally, pollution from industrial and agricultural production continues to rise. When can we expect this beneficial dynamic to take hold in other countries?

Recent data suggests that sulfur dioxide emissions even from rapidly industrializing China may have peaked in 2006 and have begun declining. Earlier studies cite evidence for a pollution turning point at which people begin to demand reductions in sulfur dioxide emissions when their per capita annual incomes reach a threshold of around $10,000 (purchasing power parity). The researchers in that study concluded, "One important lesson here is that it is possible to reduce emissions that are by-products of a modern economy, without sacrificing long-term growth."

In the face of the overwhelming evidence to the contrary, why do so many Americans still believe that air pollution is getting worse? When crime rates fall, mayors, police chiefs, and district attorneys are eager to spread the news and take the credit. But when pollution levels fall, environmentalists and environmental bureaucrats show a peculiar reluctance to cheer. The difference is that the environmental movement uses scare stories to raise money for their campaigns: no crisis, no money, no movement. In other words, Americans believe that air pollution is getting worse, as cynical as it sounds, because activists make a living peddling fear.

Doing More with Less

Jesse Ausubel, head of the Program for the Human Environment at Rockefeller University, and his colleagues point out: "If consumers dematerialize their intensity of use of goods and technicians produce the goods with a lower intensity of impact, people can grow in numbers and affluence without a proportionally greater environmental impact." In fact, that is happening. Modern economic growth is generally the result of constantly figuring out how to do more with less.

University of Manitoba natural scientist Vaclav Smil points out that modern technology enables humanity to create ever more value using less and less material. For example, the amount of energy it takes to produce goods has dropped steeply. Today it takes only 20 percent of the energy it took in 1900 to produce a ton of steel. Similarly, it now takes 70 percent less energy to make a ton of aluminum or cement and 80 percent less to synthesize nitrogen fertilizer than it did in 1900. In addition, technologist Ramez Naam shows that the amount of energy used to heat an average house in the United States is down 50 percent since 1978. The amount of energy needed to desalinate a gallon of water has plunged 90 percent since 1970. LED lights use about ten times less energy than incandescents. Humanity has gotten richer over the past couple of centuries not chiefly by doing more of the same old things, but by developing better recipes.

Excluding construction materials, Smil calculates that in the United States it once took about 10 ounces of materials back in 1920 to produce a dollar's worth of value, but that is now accomplished using only about 2.5 ounces, yielding a 75 percent decline in material intensity.

With regard to energy consumption, the American economy between 1970 and 2010 has wrung ever more value out of each kilowatt-hour and gallon of gasoline. A 2013 study by the green lobby group the Alliance to Save Energy reported, "Over the past forty years, the United States made significant gains in energy productivity. U.S. economic output expanded more than three times since 1970 while demand for energy grew only 50%." The ASE study also cited data from the energy conservation think tank the Rocky Mountain Institute suggesting that "if energy productivity had remained constant since 1970 [when about 68 quadrillion Btu (Q or quad) were consumed], the U.S. would have consumed 207.3 quadrillion Btu in 2007, when it actually only consumed 101.6 quads." A quad is roughly equivalent to 170 million barrels of oil.

While the ever more efficient use of energy and materials results in relative dematerialization—less stuff yielding more value—the overall trend has been to extract more and more materials from the earth and the biosphere. "There can be no doubt that relative dematerialization has been the key (and not infrequently the dominant) factor promoting often massive expansion of material consumption," writes Smil. "Less has thus been an enabling agent of more." For example, the 11 million cell phones in use in 1990 each bulked about 21 ounces for total overall mass of 7,000 tons. By 2011 cell phones averaged about 4 ounces, but the total weight of all 6 billion had increased a hundredfold to 700,000 tons.

As increases in efficiency make goods cheaper, people demand more of them. Initially this calculation makes it appear that people are using more resources rather than less, but this is likely wrong. Consider that billions of smartphone users living in poorer countries have skipped over the resource-intensive phase of building out millions of miles of wire phone lines, deploying tens of millions of clunky desktop computers and printers for both the home and the office, cameras and film processing, and so many other capabilities that are now embodied in four-ounce phones.

Nevertheless, Smil doubts that the current trajectory of dematerialization will speed up enough so that relative declines in material consumption translate into aggregate declines—that is, using absolutely less material while creating more value in goods and services. "The pursuit of endless growth is, obviously, an unsustainable strategy," he asserts. But what is "endless growth"? People don't want electricity, grain, housing, automobiles, and so forth. What they want is lighting, tasty food, comfortable lodging, and convenient transportation.

As a plausible scenario of how demand for materials could rise, Smil calculates that if automobile ownership in currently poor countries rises to just a third of the level in Japan (600 vehicles per 1,000 people), that would double the global fleet to 2.2 billion vehicles. However, the advent of self-driving vehicles could provide a technological end run around such projections of a growing vehicle fleet. Instead of sitting idle for most of every day, as the vast majority of automobiles do now, cars could be rented on demand.

Researchers at the University of Texas, devising a realistic simulation of vehicle use in cities that took into account issues like congestion and rush-hour usage, found that each shared autonomous vehicle could replace eleven conventional vehicles. Notionally then, it would take only about 800 million vehicles to supply all the transportation services for 9 billion people. That figure is 200 million vehicles fewer than the current world fleet of 1 billion automobiles.

In the Texas simulations, riders waited an average of 18 seconds for a driverless vehicle to show up, and each vehicle served 31 to 41 travelers per day. Less than half of 1 percent of travelers waited more than five minutes for a vehicle. In addition, shared autonomous vehicles would also cut an individual's average cost of travel by as much as 75 percent in comparison to conventional driver-owned vehicles. This could actually lead to the contraction of the world's vehicle fleet as more people forgo the costs and hassles of ownership.

In addition, a shift to fleets of autonomous vehicles makes the clean electrification of transportation much more feasible, since such automobiles could drive themselves off for recharging and cleaning during periods of low demand. Such vehicles would also be much smaller and packed more tightly on roads, since they can travel safely at higher speeds than human-driven automobiles. Such a switch would imply the construction of far less material-heavy transportation infrastructure. And fewer vehicles means that much of the 20 percent of urban land devoted to parking can be transformed into housing and businesses.

Smil worries that energy production and consumption technologies are so capital intensive that humanity will be locked into dependence on increasingly scarce and expensive fossil fuels for decades to come. Previous energy supply and consumption transformations have indeed taken decades to play out, but perhaps the energy future will follow a deployment path similar to that of information technologies.

Two decades ago, most prognosticators did not foresee how the world would skip over building landline telephone infrastructure to cellular phones. In fact, worldwide, there are in 2014 only about 1.1 billion fixed telephone landlines compared to more than 7 billion cellular phone subscriptions. I make no predictions, but increasingly cheap solar panels attached to cheap high efficiency batteries powering miserly lights, appliances, and infotech is not out of the question. Trying to forecast how much energy people living in 2100 will be using and what technologies they will be powering is like assembling a committee composed of luminaries like Thomas Edison, Madame Curie, and Albert Einstein in 1900 to accurately project how much energy we use today and how we use it.

Some trends do, in fact, indicate that humanity is withdrawing from the natural world.

In a 2014 analysis, Iddo Wernick, a researcher at Rockefeller University's Program for the Human Environment, presented data on resource consumption trends that suggests that improving efficiency and changing consumer preferences are outrunning the demands from rising population and affluence to actually reduce in many cases the amounts of material that Americans and the rest of the world use.

Wernick and his colleagues collected consumption data on a hundred materials that have long been used in the US economy. The commodities were sorted into three categories: those in which both intensity of use (kilograms per dollar of GDP) and absolute consumption (kilograms overall) are falling; those in which intensity of use is falling but absolute consumption is still increasing; and those in which both intensity of use and absolute consumption are increasing.

Thirty-six of these materials fall into the first category, including chromium, iron ore, pig iron, copper, lead, and asbestos. Fifty-three fell into the second group, among them corn, electricity, nitrogen, beef, nickel, and petroleum. Wernick believes that many of these commodities will soon reach their absolute peak—that is, the point where an economy decreases its consumption of a material resource even as economic growth and increases in wealth continue to multiply. For example, nitrogen fertilizer use has been essentially flat since the 1980s even as crop yields have risen. US population increased 80 million since 1980, yet the country uses no more water than it did then.

And then there are the eleven commodities for which both intensity of use and absolute amounts are still increasing. These include diamonds, gallium, rhenium, niobium, helium, garnets, and chicken. Wernick pointed out that while the absolute amounts of these eleven commodities are still increasing, the actual tonnage is quite small. Except for chicken, most of the commodities in this group function as technological "vitamins" that enhance the efficiency of many other industrial processes and technologies.

Why chicken? In part, because Americans are substituting it for beef. Program for the Human Environment director Jesse Ausubel outlined an input productivity hierarchy of meats, analogizing beef to getting 12 miles per gallon, pork 40 mpg, chicken 60 mpg, and tilapia and catfish 80 mpg.

How do the trends look in the rest of the world? Those data are much sparser, but Wernick was able to find reliable information in some cases. Japanese aluminum consumption, like US aluminum consumption, peaked in the 1990s. Per capita petroleum consumption peaked in the United States around 1970 and in Japan and South Korea in the 1990s. China and India are both on the early part of their consumption curves for materials, yet Wernick argues that "while Asian countries are at different stages of development, they show similar patterns of eventual saturation." Ausubel observed that Japan and Europe are paralleling materials consumption patterns identified in the United States. "I expect that in two or three decades it will be the same story in China and India," he added.

Furthermore, research by Jesse Ausubel and his colleagues suggests that humanity has reached peak farmland. Crop productivity is increasing so much that farmers will increasingly leave more and more land for nature. "The 21st century will see release of vast areas of land, hundreds of millions of hectares, more than twice the area of France for nature," declared Jesse Ausubel in 2012. In addition, requirements for synthesized nitrogen fertilizer may moderate as crop plants bioengineered to be nitrogen-sparing are deployed.

The development of lab-grown meat could well obviate Smil's advocacy of a more or less vegetarian diet in order to reduce environmentally damaging material flows. Researchers argue that cultured meat would require up to 99 percent less land, 96 percent less water, and 45 percent less energy, and would produce up to 96 percent less greenhouse gas emissions. As a proof of concept, researchers at New Harvest backed by Google founder Sergey Brin produced a lab-grown hamburger in 2013. The team is now forging onward "building a progressive food system that is sustainable, healthy and humane."

Banning Garrett, founding director of the Atlantic Council's Strategic Foresight Initiative, asserts that additive manufacturing "is likely to play a significant role in dramatically increasing the efficiency of resource use and in lowering overall carbon emissions, from the process of manufacturing and to delivering products to the end user. As only the material needed for parts is used, there is nearly zero waste." The US Department of Energy's Advanced Manufacturing Office noted, "Additive manufacturing has the potential to vastly accelerate innovation, compress supply chains, minimize materials and energy usage, and reduce waste." Additive manufacturing is also known as 3-D printing; machines build up new items one layer at a time. The Advanced Manufacturing Office suggested that additive manufacturing can reduce material needs and costs by up to 90 percent. And instead of the replacement of worn-out items, their material can simply be recycled through a printer to return it to good-as-new condition using only 2 to 25 percent of the energy required to make new parts. In addition, 3-D printing on demand will eliminate storage and inventory costs, and significantly cut transportation costs.

Sustainable Development

"The current global development model is unsustainable." That was the conclusion of the High-Level Panel on Global Sustainability, appointed in 2012 by UN secretary-general Ban Ki-moon to outline the economic and social changes needed to achieve global sustainability. The panel urged world leaders to embrace "a new approach to the political economy of sustainable development."

The panel's report, *Resilient People, Resilient Planet: A Future Worth Choosing,* specifically cited the definition of sustainable development devised in *Our Common Future,* another UN report from an expert panel headed by former Norwegian prime minister Gro Harlem Brundtland, issued in 1987. "Sustainable development is development that meets the needs of the present without compromising the ability of future generations to meet their own needs," declared the Brundtland report.

It turns out that the only form of society that has so far met this criterion is democratic free-market capitalism. How can that be? Let's take a look at the two terms, *sustainable* and *development.* With regard to most of human history, there has been precious little in the way of development. The vast majority of people lived and died in humanity's natural state of disease-ridden abject poverty and pervasive ignorance. For example, as British economic historian Angus Maddison shows, economic growth proceeded at the stately pace of less than 0.1 percent per year in Western Europe between AD 1 and 1820, rising in constant dollars from $425 in AD 1 to $1,200 in 1820. World per capita GDP rose from $467 in AD 1 to $666 in 1820.

And what about the other term, sustainable? Again, looking across history and the globe, we know for a fact that there have been until now no sustainable societies. All of the earlier civilizations in both the Old and New Worlds collapsed at various times—for example, Babylonia, Rome, the Umayyad Caliphate, Harappan, Gupta, Tang, Songhai, Mayan, Olmec, Anasazi, Moche, just to mention a few. Of course, collapse in this context doesn't mean that everybody died, but that their ways of life radically shifted and often much of the population migrated to other regions. In other words, history provides us with no models of sustainable development other than democratic capitalism.

Every one of these earlier ultimately unsustainable societies was what economics Nobelist Douglass North and his colleagues call, in *Violence and Social Orders: A Conceptual Framework for Interpreting Recorded Human History*, "natural states." Natural states are basically organized as hierarchical patron-client networks in which small, militarily potent elites extract resources from a subject population. The basic deal is a Hobbesian contract in which elites promise their subjects an end to the "war of all against all" in exchange for wealth and power.

Natural states operate by limiting access to valuable resources—that is to say, by creating and sharing the rewards of monopolies. One fundamental downside to this form of social organization is that innovation, both social and technological, is stifled because it threatens the monopolies through which elite patrons extract wealth. But why don't extractive elites encourage economic growth? After all, economic growth would mean more wealth for them to loot.

In their 2012 book *Why Nations Fail: The Origins of Power, Prosperity, and Poverty*, MIT economist Daron Acemoğlu and Harvard economist James Robinson largely concur with the analysis of North and his colleagues. They too find that since the Neolithic agricultural revolution, most societies have been organized around "extractive" political and economic institutions that funnel resources from the mass of people to small but powerful elites. The economic and political institutions that produce economic growth are inevitable threats to the power of reigning elites. "The fear of creative destruction is the main reason why there was no sustained increase in living standards between the Neolithic and Industrial revolutions. Technological innovation makes human societies prosperous, but also involves the replacement of the old with the new, and the destruction of the economic privileges and political power of certain people," they explain. Thus throughout history, reactionary elites have predictably resisted innovation because of their accurate fear that it would produce rivals for their power.

So while natural states do succeed in dramatically reducing interpersonal violence, they have one appalling consequence, as Maddison's data show: persistently low average incomes. Again, as history teaches, civilizations organized as natural states are not sustainable in the long run.

Lots of thinkers have pondered what causes the collapse of civilizations—that is to say, why they are unsustainable over the long run. Let's take a brief look at three recent theories of unsustainability: climate change, complexity, and self-organized criticality cascades. In the January 26, 2001, issue of *Science*, Yale University anthropologist Harvey Weiss and University of Massachusetts geoscientist Raymond Bradley asked, "What Drives Societal Collapse?" They concluded, "Many lines of evidence now point to climate forcing as the primary agent in repeated social collapse." Basically they argue that abrupt and longlasting droughts caused the downfall of civilizations in both the Old and New Worlds.

Utah State University anthropologist Joseph Tainter, author of the 1988 classic *The Collapse of Complex Societies,* asserts that societies fall apart when their problem-solving institutions fail. Tainter argues, "Confronted with problems, we often respond by developing more complex technologies, establishing new institutions, adding more specialists or bureaucratic levels to an institution, increasing organization or regulation, or gathering and processing more information."

Tainter maintains that this strategy of building complex institutions ultimately fails as the result of diminishing marginal returns to the social investment in them. Collapse occurs when an accumulation of unaddressed problems overwhelm a society. Interestingly, Tainter notes, "In a hierarchical institution, the flow of information from the bottom to the top is frequently inaccurate and ineffective."

In a 2002 article, "Why Do Societies Collapse?" published in the *Journal of Theoretical Politics,* independent political scientist Gregory Brunk argues that societies are self-organizing critical systems. The usual example of self-organizing criticality is a sandpile to which grains of sand are constantly being added. Many land and simply find a place in the pile; some grains land and cause small local avalanches, which soon come to rest; and eventually a grain lands that causes a huge avalanche that changes the shape of the whole pile. In a 2009 article, "Society as a Self-Organized Critical System," in *Cybernetics and Human Knowing,* researchers Thomas Kron and Thomas Grund suggest the example of the start of World War I as a social avalanche. In that case, an unlikely series of events involving a lost driver gave Serbian nationalist assassin Gavrilo Princip the opportunity to kill Franz Ferdinand, the archduke of the Austro-Hungarian Empire, and his wife, Sophie. And as the phrase goes, the rest was history.

Brunk suggests the main mechanism by which societies reach a critical point where collapses are realized was outlined by economist Mancur Olson in his 1982 book *The Rise and Decline of Nations: Economic Growth, Stagflation, and Social Rigidities.* Olson argued that over time interest group politics produces overbureaucratization, essentially re-creating the patron-client networks characteristic of natural states.

These three theories of societal collapse can complement one another. Long duration intense local droughts would no doubt constitute a problem that complex hierarchical institutions would have difficulty solving, thus producing a criticality cascade that results in social collapse. It's important to stress that all of the social collapses cited by these authors occurred in natural states—that is, societies organized as patron-client networks. In fact, the more recent social collapses—for example, the Soviet Union, Yugoslavia, the Congo, Somalia, Libya, Syria, and Iraq—all also occurred in residual natural states that had persisted into the modern era.

The plain fact is that development (rising incomes, health, and education) occurred only after what North and his colleagues identify as a new form of social organization, open-access social orders, arose during the past two centuries. Open-access social orders are basically democratic free-market capitalist societies and are characterized by the rule of law; the proliferation of private economic, social, religious, and political institutions; and civilian control of the military. In all of history, the only kind of development has been capitalist development, along with parasitical versions of development that some remaining natural states can attain for a while by imitating aspects of open-access social orders, especially by deploying their modern technologies. By 2008, average per capita income in Western Europe was $22,200 and in China $6,800.

Is free-market development sustainable? After all, it's only been around for two hundred years. Clearly, the folks on the United Nations High-Level Panel on Global Sustainability don't think so. In September 2012 a UN-sponsored activist conference issued a declaration, *Sustainable Societies, Responsive Citizens,* that urged the replacement of "the current economic model, which promotes unsustainable consumption and production patterns, facilitates a grossly inequitable trading system, fails to eradicate poverty, assists in the exploitation of natural resources to the verge of extinction and total depletion, and has induced multiple crises on Earth" with "sustainable economies in the community, local, national, regional and international spheres."

Perhaps free-market capitalism will prove itself unsustainable in the long run. But I don't think so. Brunk suggests that humans don't just take complexity cascades (avalanches) lying down; they attempt to foresee and dampen them. "From this perspective, *the fundamental reason that civilization has advanced is because societies have become more adept in addressing the problems caused by complexity cascades* [emphasis in original]," claims Brunk. The chief way in which modern societies have "become more adept in addressing the problems caused by complexity cascades" is free markets. Free markets are the most robust mechanism ever devised by humanity for delivering rapid feedback on how decisions turn out. Profits and losses discipline people to learn quickly from and fix their mistakes. Consequently, markets are superb at using trial and error to find solutions to problems.

What about the Brundtland report criterion? There is only one proven way to improve the lot of hundreds of millions of poor people, and that is democratic capitalism. It is in rich democratic capitalist countries that the air and water are becoming cleaner, forests are expanding, food is abundant, education is universal, and women's rights respected. Whatever slows down economic growth also slows down environmental improvement. By vastly increasing knowledge and pursuing technological progress, past generations met their needs and vastly increased the ability of our generation to meet our needs. We should do no less for future generations.

Top-down bureaucratization of the sort favored by many environmental activists moves societies back in the direction of natural states in which monopolies are secured and run by elites. Innovation would thus stall and the ability of people and societies to adapt rapidly to changing conditions, economic and ecological, via free markets and democratic politics would falter. "Ironically, instead of eliminating all complexity cascades, what the increasing bureaucratization of mature societies may do is increase the impact of the really big cascades when they overwhelm a society's barricades," argues Brunk. That's entirely correct.

What well-meaning activists and UN bureaucrats are trying to do is centrally plan the world's ecology. History suggests that that would work out about as well for humanity and the natural world as centrally planned economies did.

Economists Lucas Bretschger and Sjak Smulders argue that the decisive question isn't to focus directly on preserving the resources we already have. Instead, they ask: "Is it realistic to predict that knowledge accumulation is so powerful as to outweigh the physical limits of physical capital services and the limited substitution possibilities for natural resources?" In other words, can increasing scientific knowledge and technological innovation overcome the limitations to economic growth posed by the depletion of nonrenewable resources? And, according to Paul Romer, an economist and founding director of the NYU Stern Urbanization Project, the answer is yes.

Romer has observed, "Every generation has perceived the limits to growth that finite resources and undesirable side effects would pose if no new recipes or ideas were discovered. And every generation has underestimated the potential for finding new recipes and ideas. We consistently fail to grasp how many ideas remain to be discovered. The difficulty is the same one we have with compounding: possibilities do not merely add up; they multiply." While the production of some supplies of physical resources may peak, there is no sign that human creativity is about to peak.

There is one way to make sure that humanity runs out of resources—by slowing down the rate of technological progress. As it happens, lots of environmentalists advocate a policy that could in fact drastically slow down the rate of technological change—implementing the precautionary principle.

3

Never Do Anything
for the First Time

I HAVE FRIENDS WHO TOOK THE PRECAUTIONARY STEP of not having their daughter vaccinated against measles, mumps, and rubella. Based on the widely reported results of a very small study in *The Lancet*, my friends worried that vaccinations might harm their child by making her autistic. During the past two decades, study after study has found absolutely no link between vaccinations and autism. Naturally, I have plied them with information about the safety and health benefits of vaccines, but so far as I know, their daughter, now a teenager, is still unvaccinated against those childhood scourges. My friends' choice shows that taking a precautionary approach actually provides no sure guidance on what to do when it comes to the risks and benefits of modern technologies.

Never do anything for the first time. The strong version of the precautionary principle much favored by many environmentalists can largely be summarized by that maxim.

Environmentalist advocates of the principle will deny that that is what they are proposing. Instead, they claim that when it comes to evaluating technological risks they merely want society to be guided by the wisdom of the ancient aphorism "Better safe than sorry." But as we shall see, the precautionary principle as formulated by environmentalists goes much further and presumes that better safety lies in banning or restricting the development of new technologies. Consequently, implementing the strong version of the principle will instead make us "more sorry than safe," as Case Western Reserve University law professor Jonathan Adler has cogently argued. Why? The central issue is that proponents of the precautionary principle tend to focus on hypothetical dangers while generally failing to consider fully the power of new technologies to reduce risk.

The closest thing to a canonical version of the precautionary principle was devised by a group of thirty-two leading environmental activists meeting in 1998 at the Wingspread Center in Wisconsin. The Wingspread Consensus Statement on the Precautionary Principle reads:

> When an activity raises threats of harm to human health or the environment, precautionary measures should be taken even if some cause and effect relationships are not fully established scientifically.
>
> In this context the proponent of an activity, rather than the public, should bear the burden of proof.
>
> The process of applying the Precautionary Principle must be open, informed and democratic and must include potentially affected parties. It must also involve an examination of the full range of alternatives, including no action.

Why was this new principle needed? Because, the Wingspread conferees asserted, the deployment of modern technologies was spawning "unintended consequences affecting human health and the environment," and "existing environmental regulations and other decisions, particularly those based on risk assessment, have failed to protect adequately human health and the environment." As a consequence of these unintended side effects and the supposed regulatory inadequacy, the conferees insisted, "Corporations, government entities, organizations, communities, scientists and other individuals must adopt a precautionary approach to *all human endeavors* [emphasis added]." Contemplate for a moment this question: Are there any human endeavors of which some timorous person cannot assert that it raises a "threat" of harm to human health or the environment?

Unfortunately, parsing the precautionary principle is not a mere academic exercise. Versions of it have been incorporated into numerous international environmental treaties, such as the United Nations Framework Convention on Climate Change, the Cartagena Protocol on Biosafety, and the Persistent Organic Pollutants (POPs) treaty. Other renderings are explicitly integrated into European regulatory law.

Against critics of the principle, Chris Mooney, author of *The Republican War on Science*, asserts that it "is not an anti-science view, it is a policy view about how to minimize risk." That's clearly wrong. Beliefs about how much risk people should be allowed to take or to be exposed to are based on value judgments expressing moral views, not scientific facts.

Proving That Roaming Minotaurs Do Not Really Threaten Virgins

The strong version of the precautionary principle requires that the creator of a new technology or activity, rather than the public, should bear the burden of proof with regard to allaying fears about threats of harm allegedly posed by a new technology. "Assume that all projects or activities will be harmful, and therefore seek the least-harmful alternative. Shift the burden of proof—when consequences are uncertain, give the benefit of the doubt to nature, public health, and community well-being," explained Peter Montague from the Environmental Research Foundation in 2008. Boston University law professor George Annas, a prominent bioethicist who favors the precautionary principle, clearly understands that it is not a value-neutral concept. He has observed, "The truth of the matter is that whoever has the burden of proof loses."

Harvard law professor and former administrator of the Office of Information and Regulatory Affairs in the Obama administration Cass Sunstein agrees: "If the burden of proof is on the proponent of the activity or processes in question the Precautionary Principle would seem to impose a burden of proof that cannot be met." Why can't it be met? "The problem is that one cannot prove a negative," notes Mercatus Center analyst Adam Thierer. "An innovator cannot prove the absence of harm, but a critic or regulator can always prove that *some* theoretical harm exists. Consequently, putting the burden of proof on the innovator when that burden can't be met essentially means no innovation is permissible." Just because I can't prove that no minotaurs roam the woods surrounding my cabin in Virginia, that shouldn't mean that regulators can, as a precaution, ban virgins from visiting me. (Minotaurs are notoriously fond of the flesh of virgins.)

Anything New Is Guilty

Anything new is guilty until proven innocent. It's like demanding that newborn babies prove that they will never grow up to be serial killers or even just schoolyard bullies before they are allowed to leave the hospital. The point of maximum ignorance about the benefits and costs of any activity or product is before testing. If testing is not permitted and people can't gain experience using a new technology, the result amounts to never doing anything for the first time. The precautionary principle clearly is not a neutral risk management tool; it is specifically aimed at bestowing a political veto on new technologies and products onto opponents when one cannot be procured in the marketplace.

Steve Breyman, a professor in the Department of Science and Technology Studies at Rensselaer Polytechnic Institute, has made explicit how he sees the precautionary principle being wielded as a policy tool for radically reshaping modern societies. As he states, "I introduced as part of an overall green plan that included conservation and renewable energy, grass roots democracy, green taxes, defense conversion, deep cuts in military spending, bioregionalism, full cost accounting, the cessation of perverse subsidies, the adoption of green materials, designs and codes, green purchasing, pollution prevention, industrial ecology and zero emissions, etc., the precautionary principle could be an essential element of the transition to sustainability." It certainly would be good to adopt many of these policies, but his proposals have precious little to do with evaluating the threats to human health or environment allegedly posed by new technologies. Clearly, its boosters do not regard the precautionary principle as just a neutral risk analysis tool; it is a regulatory embodiment of egalitarian and communitarian moral values.

At the heart of the principle is the admonition that "precautionary measures should be taken even if some cause-and-effect relationships are not fully established scientifically." Of course, *all* scientific conclusions are subject to revision, and none are ever fully established. Since that is the case, the precautionary principle could logically apply to all conceivable activities, since their outcomes are always in some sense uncertain.

On its face, the strong version of the principle actually forestalls the acquisition of the sort of knowledge that would reveal how safe or risky a new technology or product is. In the case of genetically modified crops, two researchers pointed out, "the greatest uncertainty about their possible harmfulness existed before anybody had yet produced one. The precautionary principle would have instructed us not to proceed any further, and the data to show whether there are real risks would never have been produced. The same is true for every subsequent step in the process of introducing genetically modified plants. The precautionary principle will tell us not to proceed, because there is some threat of harm that cannot be conclusively ruled out, based on evidence from the preceding step. The precautionary principle will block the development of any technology if there is the slightest theoretical possibility of harm."

Let's parse the principle a bit more. One particularly troublesome issue is that some activities that promote human health might "raise threats of harm to the environment," and some activities that might be thought of as promoting the environment might "raise threats of harm to human health."

"The precautionary principle, for all its rhetorical appeal, is deeply incoherent," argues Cass Sunstein. "It is of course true that we should take precautions against some speculative dangers. But there are always risks on both sides of a decision; inaction can bring danger, but so can action. Precautions, in other words, themselves create risks—and hence the principle bans what it simultaneously requires."

Sunstein argues that five different common cognitive biases distort how people view precaution when considering novel risks. First, recent news about hazards come more easily to mind, distracting people's attention from other risks. Second, people attend far more credulously to worst-case scenarios, even if they are very unlikely to occur. Third, rather than risk a loss, people tend to prefer the status quo even when there is a high probability that a new activity or product will bestow significant benefits. Fourth, a common belief in a benign nature makes technological risks look more suspect. Fifth, people focus on the immediate effects of their decisions and ignore how competing risks play out over the longer run. Sadly, special interests and politicians understand and manipulate these cognitive weaknesses to inflame debates over the safety of technologies they dislike.

Precaution Is Dangerous

Sunstein poses the case of nuclear power, in which opponents claim that the principle says it should not be permitted, yet banning no-carbon nuclear power increases the risks of climate change caused by burning coal to produce electricity. This is not a hypothetical paradox. After a tsunami caused the Fukushima nuclear disaster in Japan, the German government decided to close all of its nuclear power plants. The country is now building more coal-fired electric power plants to replace them, and as a result, its emissions of carbon dioxide are increasing. Is that really the precautionary choice? On what evidence can one answer that question? It's really all about panic and political pandering. No wonder Sunstein dubbed the precautionary principle the "paralyzing principle."

Let's consider several other cases in which the precautionary principle presents paralyzing conundrums. Take the use of pesticides. Humanity has deployed them to better control disease-carrying insects such as flies, mosquitoes, and cockroaches, and to protect crops. Some studies have suggested that modern pesticides have helped decrease crop yield losses from wheat, rice, and corn by as much as 35 percent and losses from soybeans and potatoes by 43 percent. Clearly, pesticide use has significantly improved the health and nutrition of hundreds of millions of people. But some pesticides have had side effects on the environment, such as harming nontargeted species. The precautionary principle's "threats of harm to human health or the environment" standard gives no sure guidance on how to make a trade-off between human health and the protection of nonpest species.

Defenders of the precautionary principle also frequently point out that it is not really so novel, since it is already embedded in many United States regulatory schemes. As a prime example, proponents often cite as a model the drug approval process of the Food and Drug Administration, in which pharmaceutical companies who want to sell their medicines to the public must first prove that they are safe and effective. However, there is good evidence that the FDA's ever-intensifying search for safety is achieving the opposite and the agency is now killing more people than it saves.

For example, a 2010 study in the *Journal of Clinical Oncology* by researchers from the MD Anderson Cancer Center in Houston, Texas, found that the time from drug discovery to marketing increased from eight years in 1960 to twelve to fifteen years in 2010. Five years of this increase result from new regulations boosting the lengths and costs of clinical trials. The regulators aim to prevent cancer patients from dying from toxic new drugs. However, the cancer researchers calculate that the delays caused by requirements for lengthier trials have instead resulted in the loss of 300,000 patient life-years while saving only 16 life-years.

Conversely, speeding up drug approvals—using less caution—evidently saves lives. A 2005 National Bureau of Economic Research study found that, on balance, the faster FDA drug approvals made possible by new funding legislation passed in the 1990s saved far more lives than they endangered. In fact, new drugs saved up to 310,000 life-years compared to 55,000 life-years possibly lost to the side effects of drugs that were eventually withdrawn from the market. As the general counsel for the Competitive Enterprise Institute, a free-market think tank, Sam Kazman, has observed, "Whenever FDA announces its approval of a major new drug or device, the question that needs to be asked is: If this drug will start saving lives tomorrow, then how many people died yesterday waiting for the agency to act?" Precaution is killing people.

Or consider the case of subsidies and mandates for biofuels. Biofuels are supposed to protect us against the harms of climate change (by cutting greenhouse-warming carbon dioxide emissions from using fossil fuels) and to reduce the risks associated with US dependence on foreign oil. Nevertheless, the biofuel boom has resulted in farmers plowing land that once had been sequestered in conservation programs and using more fertilizer to produce corn. This has eliminated considerable swaths of wildlife habitat, and fertilizer runoff has created an extensive low-oxygen dead zone in the Gulf of Mexico. In addition, growing crops for biofuels instead of for food likely contributed to the recent hike in the global price of grains and thus increased hunger in poor countries. The precautionary principle offers no obvious counsel on the proper balance of those risks.

One of the first journalistic instances of the use of the phrase *scientific consensus* appears in the July 1, 1979, issue of *The Washington Post*, in an article on the safety of the artificial sweetener saccharin. "The real issue raised by saccharin is not whether it causes cancer (there is now a broad scientific consensus that it does) [parenthetical in original]," reported the *Post*. This belief was based on experiments in which mice dosed with huge amounts of the sweetener got bladder cancer. Based on this threatening information, the sweetener was listed as a precautionary measure in 1981 in the US National Toxicology Program's *Report on Carcinogens* as a substance reasonably anticipated to be a human carcinogen. Thirty years later, the National Cancer Institute reports that "there is no clear evidence that saccharin causes cancer in humans." In light of this new scientific consensus, the sweetener was delisted as a probable carcinogen in 2000. In this instance precaution was exercised, just as the principle admonishes, because "some cause and effect relationships are not fully established scientifically." The result was that offering this safe low-calorie sweetener in the marketplace was substantially hindered just as obesity rates in America were skyrocketing.

If pesticides and nuclear power aren't bad enough, what about the dangers posed by the source of energy that practically defines the modern age, electricity? In 1989, *New Yorker* staff writer Paul Brodeur launched the fear campaign that electromagnetic fields (EMF) generated by power lines, household appliances, television screens, and electric blankets were causing an epidemic of cancer. Following the model of Rachel Carson's *Silent Spring*, Brodeur's *New Yorker* articles became the 1993 book *The Great Power-Line Cover-up: How the Utilities and the Government Are Trying to Hide the Cancer Hazard Posed by Electromagnetic Fields*.

By 1997, the National Academy of Sciences had released a report on EMF based on a three-year review of five hundred epidemiological studies that concluded that "the current body of evidence does not show that exposure to these fields presents a human-health hazard. Specifically, no conclusive and consistent evidence shows that exposures to residential electric and magnetic fields produce cancer, adverse neurobehavioral effects, or reproductive and developmental effects." A year later, the National Cancer Institute issued a seven-year epidemiological study that found no connection whatsoever between exposure to EMFs and childhood leukemia.

As physicist Robert Park later concluded, "The EMF controversy has faded from view. But think of the damage done in the meantime. Hundreds of millions of dollars were spent in litigation between electric utilities and tort lawyers; homeowners near power lines saw a collapse in property values; municipalities had to pay for new electrical work in schools. The real cost, though, is human—millions of parents were terrified to no purpose."

Remember the cell phone cancer panic? In January 1993, David Reynard of St. Petersburg, Florida, appeared on CNN's *Larry King Live*, where he claimed that his wife died of a brain tumor that developed where she held her cellular phone to her head. Reynard filed a suit against the cell phone's maker and the phone company. Thus was launched the great cell phone cancer scare, which continues to this day. In 2010, the city of San Francisco, California, specifically citing the Wingspread version of the precautionary principle, passed an ordinance requiring radiation warning labels on mobile phones. It took until 2013 for the city to drop its labeling requirement in the face of lawsuits. The National Cancer Institute now states with regard to cell phones that "to date there is no evidence from studies of cells, animals, or humans that radiofrequency energy can cause cancer."

Consider also how the precautionary principle would have applied to the introduction of cellular phones. The principle mandates that a new technology not be permitted whenever someone alleges that it might potentially cause harm to human health. So it takes little imagination to think about what would have happened to this increasingly important means of communication had the precautionary principle been invoked as the cell phone cancer scare was spreading through the media. In 1993—when Reynard made his claims—10 million Americans were using cell phones and there were probably fewer than 20 million users around the globe. Today, there are 7 billion mobile phone subscriptions worldwide. By the way, Reynard lost his lawsuit.

New Attacks on New Technologies

The promoters of fear never rest—the modern world abounds in new threatening technologies. Consider, for example, the promising new suite of technologies that comprise synthetic biology. In 2012, a coalition of 111 environmental and social activist groups called for a moratorium on the development of synthetic biology. The activists are worried about researchers into synthetic biology who aim to create a toolbox of standardized intracellular parts that can be used to create novel organisms that do things like clean up toxic wastes, make fuels, or produce medicines. The moratorium declaration specifically cites the Wingspread Consensus Statement as its authoritative version of the precautionary principle. The declarants state, "Applying the Precautionary Principle to the field of synthetic biology first necessitates a moratorium on the release and commercial use of synthetic organisms, cells, or genomes."

Once the moratorium is in place, the groups want governments to conduct "full and inclusive assessments of the implications of this technology, including but not limited to devising a comprehensive means of assessing the human health, environmental, and socio-economic impacts of synthetic biology." It's not just risks to health and environment that are to be weighed, but also social and economic risks that are to be assessed. The Friends of the Earth letter accompanying the call for a moratorium noted that it is "rooted in the precautionary principle and the belief that the health of people and our environment must take precedence over corporate profits." For what it's worth, President Obama's Commission for the Study of Bioethical Issues issued a comprehensive report in 2010, noting that "synthetic biology does not necessarily raise radically new concerns or risks." The commission explicitly rejected applying the precautionary principle to synthetic biology and instead recommended "an ongoing system of *prudent vigilance* that carefully monitors, identifies, and mitigates potential and realized harms over time." The commission concluded that with respect to the benefits and harms of synthetic biology, the current regulatory system is robust enough to protect people and the environment.

Nanotechnology is also being targeted by proponents of the precautionary principle. Nanotechnology basically encompasses a suite of new technologies involving the use of materials at scales measuring in billionths of an inch, including tools like 3-D printing and carbon nano-tubes. When it comes to regulating nanotechnology, Georgia Miller from Friends of the Earth asks, "Who is afraid of the precautionary principle?" She argues for "a more comprehensive application of the pre-cautionary principle [that] would see nanotechnology's broader socio-economic and political implications considered and assessed alongside its toxicity risks." The proponents of the precautionary principle ulti-mately hope that blocking the development of new technologies will force the rest of us to submit to the more radically communitarian and egalitarian forms of society and economics that they prefer.

The Seen and the Unseen

Promoters of the precautionary principle argue that its great advan-tage is that implementing it will help avoid deleterious unintended consequences of new technologies. Unfortunately, supporters are most often focusing on the *seen* while ignoring the *unseen*. In his brilliant es-say "What Is Seen and What Is Unseen," nineteenth-century French economist Frédéric Bastiat pointed out that the favorable "seen" effects of any policy often produce many disastrous "unseen" later consequences. Bastiat urges us "not to judge things solely by *what is seen*, but rather by *what is not seen*." Banning nuclear power plants reduces the alleg-edly seen risk of exposure to radiation while boosting the unseen risks associated with man-made global warming. Prohibiting a pesticide aims to diminish the seen risk of cancer, but elevates the unseen risk of malaria. Demanding more drug trials seeks to prevent the seen risks of toxic side effects, but increases the unseen risks of disability and death stemming from delays in getting effective drugs to patients. Mandating the production of biofuels attempts to address the seen risks of depen-dence on foreign oil, but heightens the unseen risks of starvation.

Jonathan Adler sensibly asks, "Why is it safer or more 'precaution-ary' to focus on the potential harms of new activities or technologies without reference to the activities or technologies they might displace?"

He adds, "There is no a priori reason to assume that newer technologies or less-known risks are more dangerous than older technologies or familiar threats. In many cases, the exact opposite will be true. A new, targeted pesticide may pose fewer health and environmental risks than a pesticide developed ten, twenty, or thirty years ago. Shifting the burden of proof, as the Wingspread Statement calls for, is not a 'precautionary' policy so much as a reactionary one."

As we've seen, the precautionary principle privileges the status quo by shifting the burden of proof to the proponents of new activities and technologies. But that's not all. The rhetoric of the precautionary principle enables its promoters to drape themselves in the mantle of the public interest. The precautionary principle "places the speaker on the side of the citizen—I am acting for your health—and portrays the opponents of the contemplated ban or regulation as indifferent or hostile to the public's health," explained Aaron Wildavsky. "The rhetoric works in part because it assumes what actually should be proved, namely that the health effects of the regulation will be superior to the alternative. This comparison is made possible in the only possible way—by assuming that there are no health detriments from the proposed regulation." In other words, proponents of the precautionary principle are trying to get away with claiming that there are no trade-offs; they assert that their policy of suppressing new technologies guarantees benefits without incurring any risks. But as the pesticide, nuclear power, drug approval, and biofuels examples clearly show, that is simply not true. It is impossible to abate just one risk—there are risks on all sides of any technological process, including the risks of banning it.

Precautionists also ignore another vital fact about progress: All technologies serve as bridges to other technologies, to ever better and safer alternatives. For example, without the production of fossil fuels, humanity would not be in the position to make the costly, knowledge-intensive transition to the solar/hydrogen future that many environmentalists wish to subsidize into existence. One technology leads to another. As dirty as burning fossil fuels may be, they aren't a tenth as dirty as burning wood. And if the world had not switched to fossil fuels, it might well have been the case that all the world's forests would have been cut down by now.

Precaution and Perfect Foresight

Embedded in the precautionary principle is the notion that humans can somehow anticipate all of the ramifications of a technology in advance and can tell whether on balance it will be a net benefit or a net cost to humanity and the environment. That's complete nonsense. Human beings are terrible at foresight. To cite a single example, when the optical laser was invented in 1960, it was dismissed as "an invention looking for a job." No one could imagine what possible use this interesting phenomenon might be. Of course, now the optical laser is integral to the operation of hundreds of everyday products. It runs our printers, transmits our data on optical telephone networks, performs laser surgery to correct myopia, checks us out at the store, plays our CDs, opens clogged arteries, helps level our crop fields, and so forth. It's ubiquitous. Yet no one anticipated—no one could have anticipated—how incredibly useful lasers would turn out to be, not even the wisest tribunal of environmentalist seers. Permissionless innovation produces progress.

Consider another case of overwrought precaution. In the 1970s, there were extensive efforts to ban genetic engineering research on precautionary grounds (see the next chapter). As late as 1989, in response to Green Party pressure, German regulators forbade the chemical manufacturer Hoechst to open its then-state-of-the-art biotechnology facility outside Frankfurt to produce pure human insulin using gene-spliced bacteria. The process poses no significant threats and biotech insulin is safer for diabetics to use than was the standard pig and cow insulin sourced from slaughterhouses. The anti-biotech precautionary prohibition doubtlessly harmed German diabetics who would have benefited from nonanimal insulin.

Electricity, automobiles, antibiotics, oil production, computers, plastics, vaccinations, chlorination, mining, pesticides, paper manufacture, and nearly everything that constitutes the vast enterprise of modern technology all have risks. On the other hand, it should be perfectly obvious that allowing inventors and entrepreneurs to take those risks has enormously lessened others. How do we know? People in modern societies are enjoying much longer and healthier lives than did our ancestors, with greatly reduced risks of disease, disability, and early death.

"A generic focus on new products is problematic because they often present lower risks than the older products they are intended to replace and failing to adopt new products can increase risks," observes a 2013 report from the nonprofit Council for Agricultural Science and Technology. "Regardless of whether the subject is automobiles, pharmaceuticals, pesticides, factories, or a myriad of other products, new technologies are generally safer than the older versions. By imposing a barrier to the introduction of newer technologies, the Precautionary Principle favors the status quo, which could often mean higher risks."

The precautionary principle is profoundly conservative, privileging the old over the new, the past over the future. It amounts to an argument from ignorance, claiming that the truth of a premise is based on the fact that it has not been proven false, or that a premise is false because it has not been proven true. In this case, the wielders of the precautionary principle can simply assert that any new technology they dislike could be dangerous merely because their claim has not been proven false. Conversely, precautionists can contend that claims for the safety of a new technology are false on the grounds that it has not been proven (sufficiently) true.

Prior to the modern era, most societies were dominated by elites that sought to restrict the range of activities and technologies available to their subjects. For example, on precautionary grounds the samurai during the Tokugawa period in Japan forbade their subjects firearms; the Turkish caliph outlawed printing presses throughout the Middle East and North Africa until 1729; and the Chinese emperor burned all oceangoing vessels in 1525 and restricted ships to having just two masts for sails. Now modern environmentalist elites would similarly restrict access to technologies that they find too dangerous and socially disruptive.

The precautionary principle empowers a self-selected elite of the timorous to obstruct progress for the majority. In a sense, the precautionary principle is a return to the era when clerics and nobles (environmentalist ideologues and bureaucrats today) had the power to halt innovations on the grounds that they were bad for the common folk. The precautionary principle is the opposite of the scientific process of trial and error that is the modern engine of knowledge and prosperity. The precautionary principle impossibly demands trials without errors, successes without failures.

Trial Without Error

"The direct implication of trial without error is obvious: If you can do nothing without knowing first how it will turn out, you cannot do anything at all," explained Aaron Wildavsky in his 1988 book *Searching for Safety*. "An indirect implication of trial without error is that if trying new things is made more costly, there will be fewer departures from past practice; this very lack of change may itself be dangerous in forgoing chances to reduce existing hazards."

Wildavsky added, "Existing hazards will continue to cause harm if we fail to reduce them by taking advantage of the opportunity to benefit from repeated trials." On the other hand, he suggested, "Allowing, indeed, encouraging, trial and error should lead to many more winners, because of (a) increased wealth, (b) increased knowledge, and (c) increased coping mechanisms, i.e., increased resilience in general." Wildavsky contends that pursuing a strategy of resilience is a far superior way to mitigate any deleterious side effects of new technologies. Greater knowledge, experience, and wealth gained from technological progress supplies societies and individuals with more options for handling whatever problems might arise, either natural or man-made.

Progress and Safety Happen Only Through Trial and Error

Fortunately, two centuries ago, some societies managed to escape the dead hand of elite rule and embark upon the trial-and-error process embodied in science, the market, and democratic politics. The result of the risks taken by social, economic, political, and scientific innovators is modern prosperity. "The true key to the timing of the Industrial Revolution has to be sought in the scientific revolution of the seventeenth century and the Enlightenment movement of the eighteenth century. The key to the Industrial Revolution was technology, technology is knowledge," explains Northwestern University economic historian Joel Mokyr in his 2002 book *The Gifts of Athena: Historical Origins of the Knowledge Economy*. Technology is the productive engine that has enabled some happy portion of humanity to escape from our natural state of abject poverty.

Correspondingly, Timothy Ferris, author of *The Science of Liberty: Democracy, Reason, and the Laws of Nature*, points out: "Liberalism and science are methods, not ideologies." Both embody the freedom to explore and experiment, enabling people to more systematically use trial and error to seek truths about the physical and social worlds. Both science and liberalism advance in their goal of better understanding their subject matter by falsifying asserted claims. As Nobel Prize economics laureate Friedrich Hayek argued, "Human reason can neither predict nor deliberately shape its own future. Its advances consist in finding out where it has been wrong." It is through a continual process of trial and error and success and failure that science and liberalism ultimately yield better ways of doing things.

The modern combination of liberal trial-and-error institutions—limited democracy, free markets, and liberal science—emerged in Western Europe and North America and have been spreading around the globe. Wherever the institutions of liberalism have been embraced, prosperity has followed in their wake. Given all the benefits that modern scientific and technological enterprise has bestowed upon humanity, why would some people be against it? "Technological progress inevitably involves losers, and these losers . . . tend to be concentrated and usually find it easy to organize," notes Mokyr. "Sooner or later in any society the progress of technology will grind to a halt because the forces that used to support innovation become vested interests," he explains. "In a purely dialectical fashion, technological progress creates the very forces that eventually destroy it."

As Jonathan Adler notes, "Economic interests also have reason to adopt precautionary appeals insofar as such appeals enable these groups to erect barriers to competing technologies or firms, close markets, or otherwise use environmental regulations as a tool for rent-seeking." Candlemakers, after all, cannot be expected to hail the invention of the electric lightbulb, nor hostlers the advent of automobiles, nor canal-boat owners the building of railways, nor coal miners the development of nuclear power. Applying the precautionary approach, candlemakers will urge rejection of the competing technology, citing the dangers of electric shock; hostlers, of car crashes; canal-boat owners, of train engine smoke; and miners, of the risks of radiation. European governments eager to protect their farmers from competition have already cited the precautionary principle as justification for blocking the imports of meat from hormone-treated cows and genetically enhanced grains from the United States.

The great fear of many proponents of the precautionary principle is that if technological decisions are left to people voluntarily acting in markets, those who favor a new technology can vote yes by buying it or switching to it. They can purchase products using new synthetic materials, or foods grown using biotechnology, or energy produced by thorium reactors. Of course, those who oppose a new technology can refuse to buy or use it and its products; but, as Mokyr notes, they "have no control over what others do even if they feel it might affect them. In markets it is difficult to express a no vote."

Thus it is no surprise that the foes of various new technologies embrace the precautionary principle and urge that decisions about them be moved from the voluntary realm of markets to the domain of political mandates. Of course, they benignly characterize this move as being more "democratic." In reality, opponents of new technologies believe that they will have more luck in stopping technologies they abhor by lobbying their local congressperson or member of parliament to vote to prohibit their development.

"Activists, bureaucrats, and lawyers are hampering promising research and making it more costly," writes Mokyr. "But the achievements made possible by new useful knowledge in terms of economic well-being and human capabilities have been unlike anything experienced before by the human race. The question remains, can this advance be sustained?" That is indeed the question for the twenty-first century.

Friedrich Hayek explained that "it is because freedom means the renunciation of direct control of individual efforts that a free society can make use of so much more knowledge than the mind of the wisest ruler could comprehend." This includes those who think that they can anticipate and prevent harms stemming from the process of technological innovation.

As the history of the last two centuries has shown, Hayek was surely right when he concluded: "Nowhere is freedom more important than where our ignorance is greatest—at the boundaries of knowledge, in other words, where nobody can predict what lies a step ahead. . . . The ultimate aim of freedom is the enlargement of those capacities in which man surpasses his ancestors and to which each generation must endeavor to add its share—its share in the growth of knowledge and the gradual advance of moral and aesthetic beliefs, where no superior must be allowed to enforce one set of views of what is right or good and where only further experience can decide what should prevail. It is wherever man reaches beyond his present self, where the new emerges and assessment lies in the future, that liberty ultimately shows its value."

Unfortunately, the precautionary principle sounds sensible to many people, especially those who live in societies already replete with technology. These people have their centrally heated house in the woods; they already enjoy the freedom from want, disease, and ignorance that technology can provide. They may think they can afford the luxury of ultimate precaution. But there are billions of people who still yearn to have their lives transformed. For them, the precautionary principle is a warrant for continued poverty, not safety.

Should we look before we leap? Sure we should. But every utterance of proverbial wisdom has its counterpart, reflecting both the complexity and the variety of life's situations and the foolishness involved in applying a short list of hard rules to them. Given the manifold challenges of poverty and environmental renewal that technological progress can help us address in this century, the wiser maxim to heed is "He who hesitates is lost."

4

What Cancer Epidemic?

IN MY THIRTIES, I GREATLY REDUCED MY RISK of cancer when I visited a hypnotist in New York City. I started smoking cigarettes in college and quickly became a three- to four-pack-a-day smoker. Even when faced with the pain and sorrow of numerous lung cancer patients while working for a while as a hospital attendant in a radiation oncology department, I didn't quit. But eventually the data on just how dangerous tobacco smoking is sank in, so I tried stopping cold turkey, chewing packs of nicotine gum, and so forth. Nothing worked.

Thus early one morning in the mid-1980s, I found myself standing in front of an apartment building on New York's Upper East Side waiting for my appointment with a hypnotist who promised to end my tobacco habit. Although I had little faith in that sort of hocus-pocus, I was so afraid that it might work, I anxiously stood outside the building chain-puffing on what I feared would be my last cigarettes. It turned out that they were. The hypnotism worked, but I suspect that the embarrassment I would have suffered had I continued to smoke after I had told all my friends that I was spending $400 on the hypnosis session might have helped, too. The largest known percentage (30 percent) of US cancer deaths is due to tobacco. Epidemiology suggests that because I stopped smoking in my thirties, my chances of dying of lung cancer before age seventy-five are about double those of someone who never smoked, but they are considerably below the twentyfold higher risk of smokers who never quit.

Activists such as the folks at the Pesticide Action Network frequently claim that Americans are in the midst of a "cancer epidemic" and vaguely assert that there is a "growing scientific consensus that environmental contaminants are causing cancer in humans." What contaminants? In 2009, *Wall Street Journal* MarketWatch columnist Paul Farrell pronounced the litany: "Consider the deadly impact of insecticides, pesticides, herbicides, detergents, plastics . . . the list is endless."

In 2010, the prestigious President's Cancer Panel furthered promoted alarm in *Reducing Environmental Cancer Risk: What We Can Do Now*. The press release heralding the report noted "the growing body of evidence linking environmental exposures to cancer in recent years" and asserted that "the true burden of environmentally-induced cancer is greatly underestimated." By environmental exposures, the two-member panel, consisting of Howard University surgeon Dr. LaSalle D. Leffall Jr. and MD Anderson Cancer Center immunologist Margaret Kripke, largely meant man-made chemicals like plastics and pesticides.

The panel's report additionally asserted, "With nearly 80,000 chemicals on the market in the United States, many of which are used by millions of Americans in their daily lives and are un- or understudied and largely unregulated, exposure to potential environmental carcinogens is widespread." In a statement Leffall also declared, "The increasing number of known or suspected environmental carcinogens compels us to action, even though we may currently lack irrefutable proof of harm." Leffall was doing nothing less than invoking the precautionary principle. In other words, we don't know, but let's ban something anyway.

The panelists are not alone in their alarm. In the prominent journal *The Proceedings of the Royal Society B* in 2013, Paul and Anne Ehrlich warned, "Another possible threat to the continuation of civilization is global toxification. Adverse symptoms of exposure to synthetic chemicals are making some scientists increasingly nervous about effects on the human population." The Ehrlichs added that the danger of global toxification "has been clear since the days of Carson, exposing the human population to myriad subtle poisons."

Given advocacy and reports like these, is it any wonder that a 2007 American Cancer Society poll found that seven out ten Americans believed that the risk of dying of cancer is going up?

No Growing Cancer Epidemic

There's only one problem—there is no growing cancer epidemic. As the number of man-made chemicals has proliferated, your chances of dying of the disease have been dropping for more than four decades. In fact, not only have cancer death rates been declining significantly, age-adjusted cancer *incidence* rates have been falling for nearly two decades. That is, of the number of Americans in nearly any age group, fewer are actually coming down with cancer. It is good news that modern medicine has increased the five-year survival rates of cancer patients from 50 percent in the 1970s to 68 percent today. What is more remarkable is that the incidence of cancer has been falling about 0.6 percent per year since 1994. That may not sound like much, but as Dr. John Seffrin, CEO of the American Cancer Society, explains, "Because the rate continues to drop, it means that in recent years, about 100,000 people each year who would have died had cancer rates not declined are living to celebrate another birthday." Director

of the National Cancer Institute Harold Varmus also noted, "It is gratifying to see the continued steady decline in overall cancer incidence
and death rates in the United States—the result of improved methods
for preventing, detecting, and treating several types of cancer."

How did it come to be the conventional wisdom that man-made
chemicals are especially toxic and the chief sources of a modern cancer epidemic? It all began with the Carson mentioned by the Ehrlichs
in 2013—that is, Rachel Carson, the author of *Silent Spring*.

Rachel Carson Launches Modern Political Environmentalism

The modern environmentalist movement was launched at the beginning of June 1962 when *The New Yorker* published excerpts from what
would become Carson's all-out attack on synthetic chemicals, especially
the pesticide DDT. "Without this book, the environmental movement
might have been long delayed or never have developed at all," declared
then Vice President Al Gore in his introduction to the 1994 edition. The
foreword to the twenty-fifth anniversary edition accurately declared that
this book "led to environmental legislation at every level of government."

In 1999 *Time* named Carson one of the 100 People of the Century.
Seven years earlier, a panel of distinguished Americans had selected
Silent Spring as the most influential book of the previous fifty years.
"She was the very first person to knock some of the shine off of moder-
nity," asserted environmentalist Bill McKibben in a *New York Times
Magazine* article celebrating the book's fiftieth anniversary.

In *Silent Spring*, Carson crafted an ardent denunciation of modern
technology that drives environmentalist ideology today. At its heart is
this belief: Nature is beneficent, stable, and even a source of moral
good; humanity is arrogant, heedless, and often the source of moral evil.
Rachel Carson, more than any other person, is responsible for the
politicization of science that afflicts our public policy debates today.

Carson worked for years at the US Fish and Wildlife Service, eventually becoming the chief editor of that agency's publications. She achieved financial independence in the 1950s with the publication of her popular celebrations of marine ecosystems, *The Sea Around Us* and *The Edge of the Sea*. Rereading *Silent Spring* reminds one that the book's effectiveness was due mainly to Carson's passionate, poetic language describing the alleged horrors that modern synthetic chemicals visit upon defenseless nature and hapless humanity. Carson was moved to write *Silent Spring* by her increasing concern about the effects of pesticides on wildlife. Her chief villain was the pesticide DDT.

Today, the Pesticide Action Network explains, "Carson used DDT to tell the broader story of the disastrous consequences of the overuse of insecticides, and raised enough concern from her testimony before Congress to trigger the establishment of the Environmental Protection Agency (EPA)." As noted previously, fifty years later Carson has many modern disciples who continue to preach that exposure to trace amounts of synthetic chemicals are responsible for an epidemic of cancer.

The 1950s saw the advent of an array of synthetic pesticides that were hailed as modern miracles in the war against pests and weeds. First and foremost of these pest control chemicals was DDT. DDT's insecticidal properties were discovered in the late 1930s by Paul Müller, a chemist at the Swiss chemical firm J. R. Geigy. The American military started testing it in 1942, and soon the insecticide was being sprayed in war zones to protect American troops against insect-borne diseases such as typhus and malaria. In 1943 DDT famously stopped a typhus epidemic in Naples in its tracks shortly after the Allies invaded Italy. DDT was hailed as the "wonder insecticide of World War II."

After the war, American consumers and farmers quickly adopted the wonder insecticide, replacing the old-fashioned arsenic-based pesticides, which were truly nasty. Testing by the US Public Health Service and the Food and Drug Administration's Division of Pharmacology found no serious human toxicity problems with DDT. Müller, DDT's inventor, was awarded the Nobel Prize in 1948.

DDT was soon widely deployed by public health officials, who banished malaria from the southern United States with its help. The

World Health Organization credits DDT with saving 50 million to 100 million lives by preventing malaria. In 1943 Venezuela had 8,171,115 cases of malaria; by 1958, after the use of DDT, the number was down to 800. India, which had over 10 million cases of malaria in 1935, had 285,962 in 1969. In Italy the number of malaria cases dropped from 411,602 in 1945 to only 37 in 1968. In 1970, a report by the National Academy of Sciences stated that "to only a few chemicals does man owe as great a debt as to DDT."

Chemical Agriculture

The tone of a *Scientific American* article by Francis Joseph Weiss celebrating the advent of "Chemical Agriculture" was typical of much of the reporting in the early 1950s. While "[i]n 1820 about 72 per cent of the population worked in agriculture, the proportion in 1950 was only about 15 per cent," reported Weiss. "Chemical agriculture, still in its infancy, should eventually advance our agricultural efficiency at least as much as machines have in the past 150 years." This improvement in agricultural efficiency would happen because "farming is being revolutionized by new fertilizers, insecticides, fungicides, weed killers, leaf removers, soil conditioners, plant hormones, trace minerals, antibiotics and synthetic milk for pigs."

In 1952 insects, weeds, and disease cost farmers $13 billion ($115 billion in today's dollars) in crops annually. Since gross annual agricultural output at that time totaled $31 billion ($276 billion in today's dollars), it was estimated that preventing this damage by using pesticides would boost food and fiber production by 42 percent. Agricultural productivity in the United States, spurred by improvements in farming practices and technologies, has continued its exponential increase. In the second decade of the twenty-first century, US crop yield was 360 percent higher and farmers produced 262 percent more food using 2 percent less inputs like labor, seeds, fertilizer, and feed than they did in 1950. As a result, the percentage of Americans living and working on farms has dropped from 15 percent in 1950 to under 2 percent today.

But DDT and other pesticides had a dark side. They not only killed the pests at which they were aimed but sometimes killed beneficial creatures as well. The scientific controversy over the effects of DDT on wildlife, especially birds, still vexes researchers. In the late 1960s, some researchers concluded that exposure to DDT caused eggshell thinning in some bird species, especially raptors such as eagles and peregrine falcons. Thinner shells meant fewer hatchlings and declining numbers. But researchers also found that other bird species, such as quail, pheasants, and chickens, were unaffected even by large doses of DDT. Carson, the passionate defender of wildlife, was determined to spotlight these harms. Memorably, she painted a scenario in which birds had all been poisoned by insecticides, resulting in a "silent spring" in which "no birds sing."

First, let's acknowledge that Carson was right about some of the harms that extensive modern pesticide use could and did cause. Carson was correct that the popular pesticide DDT did disrupt reproduction in some raptor species. It is also the case that insect pests over time do develop resistance to pesticides, making them eventually less useful in preventing the spread of insect-borne diseases and protecting crops. In fact, the first cases of evolving insect resistance were identified in California orchards at the beginning of the twentieth century, when species of scale insects became resistant to the primitive insecticides lime sulfur and hydrogen cyanide. By 1960, 137 species of insects had developed resistance to DDT. To preserve their usefulness, pesticides clearly needed to be more judiciously deployed.

To her discredit, however, Carson largely ignored the great good modern pesticides had done, especially in protecting human health by controlling insect-borne scourges such as typhus and malaria. Barely ten years before the publication of *Silent Spring* the then-new Centers for Disease Control had finally eradicated malaria in the southeastern region of the United States in 1951. Spraying with DDT to control mosquitoes has been central to the CDC's success. Unfortunately, Carson's unwillingness to fairly balance the costs and benefits of new technologies would become a hallmark of the modern environmental movement.

Rachel Carson: Cancer Scaremonger

As a polemicist, Carson realized that tales of empty birds' nests and warnings about pesticide-resistant bugs and weeds were not enough to spur most people to fear the chemicals she opposed. The 1958 passage by Congress of the Delaney Clause, which forbade the addition of any amount of chemicals suspected of causing cancer to food, likely focused Carson's attention on that disease.

For the previous half century some researchers had been trying to prove that cancer was caused by chemical contaminants in the environment. Wilhelm Hueper, chief of environmental cancer research at the National Cancer Institute and one of the leading researchers in this area, became the major source for Carson. "Dr. Hueper now gives DDT a definite rating as a 'chemical carcinogen,'" according to Carson. After reviewing the extensive epidemiological and experimental literature, the highly precautionary International Agency for Research on Cancer has now determined that DDT is as carcinogenic as coffee. Oddly, Hueper was so blinkered by his belief that trace exposures to synthetic chemicals were a major cause of cancer in humans that he largely dismissed the notion that smoking cigarettes increased the risk of cancer.

The assertion that pesticides were dangerous human carcinogens was a stroke of public relations genius. Even people who do not care much about wildlife care a lot about their own health and the health of their children.

In 1955 the American Cancer Society (ACS) predicted that "cancer will strike one in every four Americans rather than the present estimate of one in five." The ACS straightforwardly attributed the increase to "the growing number of older persons in the population." The ACS did note that the incidence of lung cancer was increasing very rapidly, rising in the previous two decades by more than 200 percent for women and by 600 percent for men. But the ACS also noted that lung cancer "is the only form of cancer which shows so definite a tendency." Seven years later, Rachel Carson would cannily entitle her chapter on cancer "One in Four."

William Souder, author of a new biography of Carson, *On a Farther Shore*, also notes that Carson's fight against pesticides became personal. "In 1960, at the halfway point in writing *Silent Spring*, just as she was exploring the connection between pesticide exposure and human cancer, Carson was herself stricken with breast cancer." Given the relatively primitive state of medicine in the 1950s, few diseases were scarier than cancer. And deaths from cancer had been rising steeply. Carson cited government statistics showing that cancer deaths had dramatically increased from 4 percent of all deaths in 1900 to 15 percent in 1958.

"The problem that concerns us here is whether any of the chemicals we are using in our attempts to control nature play a direct or indirect role as causes of cancer," wrote Carson. Her conclusion was that "the evidence is circumstantial" but "nonetheless impressive." She added the claim that in contrast with disease germs, "man *has* put the vast majority of carcinogens into the environment." She noted that the first human exposures to DDT and other pesticides were barely more than a decade in the past. It takes time for cancer to fester, so she ominously warned, "The full maturing of whatever seeds of malignancy have been sown by these chemicals is yet to come." She further warned that we were "living in a sea of carcinogens."

Even though Carson vaguely acknowledged the paucity of evidence that man-made chemicals like pesticides were actually causing cancer, she was clearly urging policymakers and the public to take what would now be called precautionary action.

Before the decade of the 1960s was over, activists like population doomster Paul Ehrlich would imaginatively transform and heighten the fears of synthetic chemicals initially spawned by Carson. In his luridly dystopian 1969 "Eco-Catastrophe" article, Ehrlich posited that by 1973

US federal government health officials would estimate "that Americans born since 1946 (when DDT usage began) now had a life expectancy of only 49 years, and predicted that if current patterns continued, this expectancy would reach 42 years by 1980, when it might level out."

In any case, hinting at cancer doom decades away was not frightening enough. Carson cherry-picked cases in an effort to show that pesticides could wreak their carcinogenic havoc much sooner rather than later. For evidence she cited various anecdotes, including one about a woman "who abhorred spiders" and who sprayed her basement with DDT in mid-August. The woman died of acute leukemia a couple of months later. In another passage, Carson cites a man embarrassed by his roach-infested office who again sprayed DDT and who "within a short time . . . began to bruise and bleed." He was within a month of spraying diagnosed with aplastic anemia. Today cancer specialists would dismiss out of hand the implied claims that these patients' cancers could be traced to such specific pesticide exposures.

To bolster these frightening anecdotes, Carson cited data that deaths from leukemia had increased from 11.1 per 100,000 in 1950 to 14.1 in 1960. Leukemia mortality rose with pesticide use; very suspicious, no? "What does it mean? To what lethal agent or agents, new to our environment, are people now exposed with increasing frequency?" asked Carson. Fifty years later the death rate from leukemia is 7.1 per 100,000—half of what Carson cited in *Silent Spring*. In fact, the incidence rate is now 13 per 100,000.

Carson surely must have known that cancer is a disease in which the risk goes up as people age. And thanks to vaccines and new antibiotics, Americans in the 1950s were living much longer, long enough to get and die of cancer. In 1900 average life expectancy was forty-seven, and the annual death rate was 1,700 out of 100,000 Americans. By 1960, life expectancy had risen to nearly seventy years, and the annual death rate had fallen to 950 per 100,000 people. Currently, life expectancy is more than seventy-eight years, and the annual death rate is 790 per 100,000 people. Today, although only about 13 percent of Americans are over age sixty-five, they account for 53 percent of new cancer diagnoses and 69 percent of cancer deaths. Another way to think about the relationship of cancer incidence to increasing age is to note that the incidence per 100,000 men is 60 for men between ages twenty and thirty-nine, 552 for men between ages forty and sixty-four, and 2,893 for men over age sixty-five. The incidence figures per 100,000 women are 89 for women whose age is between twenty and thirty-nine, 555 for those aged between forty and sixty-four, and 1,707 for women over age sixty-five.

Carson realized that even if people didn't worry much about their own health, they did really care about that of their kids. So to ratchet up the fear factor even more, she asserted that children were especially vulnerable to the carcinogenic effects of synthetic chemicals. "The situation with respect to children is even more deeply disturbing," she wrote. "A quarter century ago, cancer in children was considered a medical rarity. *Today, more American school children die of cancer than from any other disease* [her emphasis]." In support of this claim, Carson reported that "twelve per cent of all deaths in children between the ages of one and fourteen are caused by cancer."

Although it sounds alarming, Carson's statistic is essentially meaningless unless it's given some context, which she failed to supply. It turns out that the percentage of children dying of cancer was rising because other causes of death, such as infectious diseases, were drastically declining. The American Cancer Society reports that about 10,450 children in the United States will be diagnosed with cancer in 2014 and that childhood cancers make up less than 1 percent of all cancers diagnosed each year. Childhood cancer incidence has been rising slowly over the past couple of decades at a rate of 0.6 percent per year. Consequently, the incidence rate increased from 13 per 100,000 in the 1970s to 16 per 100,000 now. There is no known cause for this slight increase. The good news is that 80 percent of kids with cancer now survive five years or more, up from 50 percent in the 1970s.

Cancer Incidence Rates Are Falling

Did cancer doom ever arrive? No. In fact, cancer incidence rate fell. According to the Centers for Disease Control and Prevention, age-adjusted incidence rates have been dropping for nearly two decades. Why? Largely because fewer Americans are smoking, more are having colonoscopies in which polyps that might become cancerous are removed, and in the early 2000s many women stopped hormone replacement therapy. With regard to hormone replacement therapy, researchers have now concluded it moderately increases the risk of breast cancer.

Back in the early 1990s, based on sketchy research, environmentalists began pushing the hypothesis that past exposure to organochlorine pesticides, such as DDT, was fueling a breast cancer epidemic. However, after years of research a major review article in 2002 in the journal *CA: A Cancer Journal for Clinicians* reported that exposure of organochlorine compounds "is not believed to be causally related to breast cancer."

With regard to overall cancer risks posed by synthetic chemicals, the American Cancer Society in its 2014 *Cancer Facts and Figures* report on cancer trends concludes: "Exposure to carcinogenic agents in occupational, community, and other settings is thought to account for a relatively small percentage of cancer deaths—about 4 percent from occupational exposures and 2 percent from environmental pollutants (man-made and naturally occurring)."

Similarly, the British organization Cancer Research UK observes that for most people "harmful chemicals and pollution pose a very minor risk." How minor? Cancer Research UK notes, "Large organizations like the World Health Organization and the International Agency for Research into Cancer have estimated that pollution and chemicals in our environment only account for about 3 percent of all cancers. Most of these cases are in people who work in certain industries and are exposed to high levels of chemicals in their jobs." Like the American Cancer Society, Cancer Research UK advises, "Lifestyle factors such as smoking, alcohol, obesity, unhealthy diets, inactivity, and heavy sun exposure account for a much larger proportion of cancers."

These statements by the American Cancer Society and Cancer Research UK mirror the findings of *Carcinogens and Anticarcinogens in the Human Diet*, the definitive 1996 report from the National Academy of Sciences. The NAS concluded that levels of both synthetic and natural carcinogens are "so low that they are unlikely to pose an appreciable cancer risk." Worse yet from the point of view of anti-chemical crusaders, the NAS added that Mother Nature's own chemicals probably cause more cancer than anything mankind has dreamed up: "Natural components of the diet may prove to be of greater concern than synthetic components with respect to cancer risk."

In *Silent Spring* Carson cites data showing that American farmers were then applying about 637 million pounds of pesticides to their crops. The most recent Environmental Protection Agency estimate is that farmers used 1.1 billion pounds in 2007. (The amount of insecticide applied to crops has been falling recently, as farmers adopt genetically enhanced insect-resistant crop varieties.)

What factors really do increase cancer risk? Smoking, drinking too much alcohol, sunburns, and eating too much food. In fact, while overall cancer incidence has been falling, cancers related to obesity—that is to say, pancreatic, liver, and kidney cancers—have risen slightly.

The DDT Ban: Environmentalism's First Triumph

The first notable triumph of modern environmentalism occurred in 1972—the banning of the pesticide that Carson so abhorred, DDT.

In 1967, the activist group the Environmental Defense Fund began bringing lawsuits against the manufacturers of DDT and agitating for bans on the pesticide in various state legislatures. By 1971, the first administrator of the newly created Environmental Protection Agency, William Ruckelshaus, ordered a hearing to look into the claims against DDT. After seven months of hearings, which produced 9,362 pages of testimony by 125 witnesses, EPA judge Edmund Sweeney ruled, "DDT is not a carcinogenic hazard to man . . . is not a mutagenic or teratogenic hazard to man . . . [and the] use of DDT under the regulations involved here [does] not have a deleterious effect on freshwater fish, estuarine organisms, wild birds or other wildlife."

But EPA administrator Ruckelshaus overruled Sweeney and banned DDT on January 1, 1972. Later Ruckelshaus would justify his actions by declaring, "The ultimate judgment [on DDT] remains political. Decisions by the government involving the use of toxic substances are political with a small 'p.' In the case of pesticides in our country, the power to make this judgment has been delegated to the Administrator of the Environmental Protection Agency." Ruckelshaus also noted in his decision that "Public concern over the widespread use of pesticides was stirred by Rachel Carson's book, *Silent Spring.*" Unfortunately, Ruckelshaus's decision set a deleterious precedent: politics and panic have figured hugely in environmentalist campaigns and regulation ever since.

Carson described the choice humanity faced as a fork in the road to the future. "The road we have long been traveling is deceptively easy, a smooth superhighway on which we progress at great speed, but at its end lies disaster," she declared. "The other fork of the road—the one 'less traveled by'—offers our last, our only chance to reach a destination that assures the preservation of our earth." This kind of apocalyptic rhetoric is now standard in today's policy debates. In any case, the opposition to *Silent Spring* arose not just because Carson was attacking the self-interests of certain corporations (which she certainly was), but also because it was clear that her larger concern was to rein in technological progress and the economic growth it fuels.

Through *Silent Spring,* Carson provided those who are alienated by modern technological progress with a model of how to wield ostensibly scientific arguments on behalf of policies and results that they prefer for other reasons. "The hostile reaction to *Silent Spring* contained the seeds of a partisan divide over environmental matters that has since hardened into a permanent wall of bitterness and mistrust," writes Souder. He adds, "There is no objective reason why environmentalism should be the exclusive province of any one political party or ideology." That conclusion is flatly wrong.

It is this legacy of public policy confirmation bias that Yale law professor Dan Kahan and his research colleagues are probing at the Yale Cultural Cognition Project. In a recent study concerning how Americans perceive climate change risk published in *Nature Climate Change*, Kahan and his colleagues find that people listen to information that reinforces their values and ignore that which does not. They observe that people who are broadly identified as being on the political left "tend to be morally suspicious of commerce and industry, to which they attribute social inequity. They therefore find it congenial to believe those forms of behavior are dangerous and worthy of restriction." On the other hand, those broadly considered as being on the political right are proponents of technological progress who worry about "collective interference with the decisions of individuals" and "tend to be skeptical of environmental risks. Such people intuitively perceive that widespread acceptance of such risks would license restrictions on commerce and industry."

As trust in other sources of authority—politicians, preachers, business leaders—has withered over the past fifty years, policy partisans are increasingly seeking to cloak their arguments in the mantle of objective science. However, the Yale researchers find that greater scientific literacy actually produces greater political polarization. As Kahan and his fellow researchers report, "For ordinary citizens, the reward for acquiring greater scientific knowledge and more reliable technical-reasoning capacities is a greater facility to discover and use—or explain away—evidence relating to their groups' positions." In other words, in policy debates, scientific claims are used to vindicate partisan values, not to reach an agreement about what is actually the case. This sort of motivated reasoning applies to partisans of both the political left and right, and both learned it from Rachel Carson.

DDT and Birds

Rachel Carson heard and cited anecdotal reports of various birds either dying of acute DDT poisoning or experiencing reproductive problems, thus giving birth to her title conceit. Her book was a popular phenomenon, and not surprisingly her claims drew the attention of a lot of researchers.

The situation regarding the harm that DDT caused some bird species is not as straightforward as one might like. Science always deals with provisional conclusions. The main hypothesis is that exposure to DDT thinned the eggshells of birds, causing them to break before hatching. Not much scientific work has been done on eggshell thinning for the past three decades, so most of the relevant scientific literature is quite dated. The significant articles were published before 1980. "[The issue] kind of died out. There's a general lack of interest," agreed Daniel W. Anderson. Anderson, at the Department of Wildlife, Fish, and Conservation Biology at the University of California at Davis, was one of the original researchers on eggshell thinning. He blames the lack of new research on a lack of funding. Besides, Anderson observes, "The questions about eggshell thinning were pretty well answered, so people moved onto other things."

The DDT/eggshell thinning bandwagon got really rolling with two scientific articles. The first study, "Decrease in Eggshell Weight in Certain Birds of Prey," by British Nature Conservancy researcher D. A. Ratcliffe, was published in *Nature* on July 8, 1967. Ratcliffe claimed that the incidence of broken eggs in the nests of peregrine falcons, sparrowhawks, and golden eagles had increased considerably since 1950. He compared eggshells collected before 1946 with eggshells collected afterward and found that the eggshells of post-1946 peregrine falcon weighed 19 percent less; those of sparrowhawks weighed 24 percent less; and those of golden eagles 8 percent less. Ratcliffe dismissed lack of food and radioactive contamination as explanations for the thinning, but noted "some physiological change evidently followed a widespread and pervasive environmental change around 1945–1947. . . . For the species examined, frequency of egg-breakage, scale of decrease in eggshell weight, subsequent status of breeding population, and exposure to persistent organic pesticides are correlated. The possibility that these phenomena are links in a causal chain is being investigated," he concluded.

Those British results were soon bolstered by the study "Chlorinated Hydrocarbons and Eggshell Changes in Raptorial and Fish-Eating Birds," published in an October 1968 issue of *Science*, and authored by Daniel Anderson and Joseph Hickey, both at the University of Wisconsin. "Catastrophic declines of three raptorial species in the United States have been accompanied by decreases in eggshell thickness that began in 1947, and have amounted to 19 percent or more, and were identical to phenomena found in Britain," they declared. The three species were peregrine falcons, bald eagles, and ospreys. They claimed that the eggshell thinning coincided with the introduction of chlorinated hydrocarbon pesticides such as DDT, and concluded that these compounds were harming certain species of birds at the tops of contaminated food chains.

Still, the researchers just had a correlation between DDT and eggshell thinning. So some did what good scientists should do—they experimented. Joel Bitman at the US Department of Agriculture fed Japanese quail a diet laced with DDT. His study, "DDT Induces a Decrease in Eggshell Calcium," published in *Nature* on October 4, 1969, found that the quail dosed with DDT had eggshells that were about 10 percent thinner than those of undosed quail. However, Bitman's findings were eventually overturned because he had also fed his quail a low-calcium diet. When the quail were fed normal amounts of calcium, the thinning effect disappeared. Studies published in *Poultry Science* found chicken eggs almost completely unaffected by high dosages of DDT.

It's not DDT per se that is thought to do the damage to eggshells, but a DDT metabolite known as DDE. Thus the most persuasive feeding study refers to it: "DDE-Induced Eggshell Thinning in the American Kestrel: A Comparison of the Field Situation and Laboratory Results." This groundbreaking study was published in the *Journal of Applied Ecology* by Jeffrey Lincer in 1975.

Kestrels, commonly called sparrowhawks, are small falcons. Lincer noted that the "inverse correlation between DDE in North American raptor eggs and eggshell thickness is clear but does not prove a causal relationship since other chemicals or factors could be involved." So to find out what effect DDE might have, Lincer fed captive kestrels a DDE-laced diet and then compared their eggs with those taken from the nests of wild kestrels. Lincer found that dietary levels of 3, 6, and 10 parts per million (ppm) of DDE resulted in eggshells that were 14 percent, 17.4 percent, and 21.7 percent thinner respectively. "Despite the recent controversy, there can be little doubt now as to the causal relationship between the global contaminant DDE and the observed eggshell thinning and the consequent population declines in several birds of prey," concluded Lincer.

Still, there is a piece missing in the full scientific picture. Despite considerable research, no one has ever identified the physiological mechanism(s) by which DDE causes eggshell thinning, according to Anderson.

There is another possibly confounding issue as well. In 1998, Royal Society for the Protection of Birds researcher Rhys Green published a study in the *Proceedings of the Royal Society B* that found that eggshell thinning of some bird species had begun fifty years before the introduction of DDT.

On the contrary side are studies that showed that DDT did not cause eggshell thinning in chickens and Japanese quail. Anderson agrees that the evidence shows that gallinaceous birds (poultry and fowls), herring gulls, and most passerine birds "aren't as sensitive to DDE as raptors." More than half of all bird species are passerine or perching birds, including crows, robins, and sparrows. But even though chickens and quail fed very high concentrations of DDE and an adequate amount of food experienced essentially no eggshell thinning or other reproductive problems, science shows pretty conclusively that it's another story for raptors.

Anderson notes that DDT and DDE levels in nature have been falling for decades. Populations of bald eagles, peregrine falcons, ospreys, and brown pelicans have all bounced back. In 1969, researchers reported finding total DDT accumulations ranging from 5,000 ppm to 2,600 ppm in the fat of North American peregrine falcons. Today, one would typically find 50 ppm in raptors, according to Anderson. Such body burdens would yield only about 2.5 ppm in eggs. Anderson notes that there appears to be a threshold of 1 to 3 ppm for DDE in eggs below which there is no eggshell thinning in even sensitive bird species. Dusting DDT on the walls of houses in developing countries to control for mosquitoes seems unlikely to cross that threshold for birds.

The Pesticide Fight Today

The environmentalists won the fight about DDT in America, so why is it still a sensitive political issue today? The main reason is the continuing fight to save millions of people from malaria. Whatever it does to different types of eggshells, DDT remains unquestionably one of the most effective ways to control the mosquitoes that carry the malaria parasite. Still, international environmentalists have instituted through the UN strict controls on DDT, with an eye on an eventual permanent ban.

In *Silent Spring*, Rachel Carson asked, "Who has decided—who has the *right* to decide—for the countless legions of people who were not consulted that the supreme value is a world without insects, even though it be also a sterile world ungraced by the curving wing of a bird in flight? The decision is that of the authoritarian temporarily entrusted with power."

Banning DDT saved thousands of raptors over the past thirty years, but outright bans and misguided fears about the pesticide cost the lives of millions of people who died of insect-borne diseases like malaria. The 200 million people who come down with malaria and the 600,000 who die of the disease every year might well wonder what authoritarian made that decision.

DDT and Breast Cancer

In 2002, National Cancer Institute (NCI) researchers, in the most exhaustive study of its kind, could find no link between increased breast cancer rates and exposure to chemicals such as the pesticides DDT and chlordane or PCBs used as coolants in electrical transformers. This study is one more in a long line that can find no link between breast cancer and exposure to synthetic chemicals.

Human brains are adapted to be pattern recognition machines. Being able to recognize a tiger's stripes in a sun-dappled bamboo forest enhances one's chances of escaping its jaws. But it turns out that we over-recognize patterns. Our brains can find patterns in anything, which is why tarot cards and astrology remain popular. Researchers have shown that people will consistently claim to identify patterns in tables of randomly generated numbers. Just as our brains succumb to optical illusions, they fall victim to causality illusions.

Another point to keep in mind is that low-probability events do occur. Even if there is only a one in a million chance of something occurring to someone, with 7 billion people on the earth, it will occur 7,000 times somewhere. As the Internet reaches its tentacles further into human society, these odd occurrences have ever greater chances of being marveled at by wider and wider audiences.

From a public policy perspective, one of the most common and problematic misperceived patterns is cancer clusters. Every year we are treated to reports of communities that purportedly suffer more than their share of cancer. We all know the script by heart: Ten people in a small town are diagnosed with cancer—say, leukemia—within five years. Victims, reporters, regulators, and trial lawyers frantically search for a corporation manufacturing some allegedly toxic chemical nearby on which to pin the blame. The victims' suffering, it's assumed, must be the result of corporate greed.

This is exactly the script that played out on Long Island back in the 1990s. Under orders from Congress, the NCI conducted the most searching inquiry ever into an alleged cancer cluster and came up with exactly nothing. The NCI's Deborah Winn told *The New York Times* the data "were very, very conclusive" that the synthetic chemicals studied "are not associated with breast cancer."

Interestingly, another study, reported just a month earlier, did find a very high correlation between a risk factor and a woman's chances of getting breast cancer. That study confirmed that the longer a woman breast-fed, the lower her risk of getting breast cancer. Could it be that long-term breast-feeding was not fashionable among well-off suburban moms on Long Island two or three decades ago?

Fate or bad luck are not acceptable to us pattern-searching humans. What we once blamed on the malevolence of witchcraft, we now blame on the malevolence of corporations. In a sense, we are still in hot pursuit of witches.

We Are Now Living Long Enough to Get Cancer

As our deepening knowledge of biology is revealing, many of the human body's resources are aimed at keeping our cells' natural tendency to become cancerous at bay long enough for us to reproduce successfully. The plain, unavoidable fact of life is that our bodies' defenses against cancer break down as we age. It is true that high, prolonged exposures to some synthetic chemicals (or, in the case of cigarettes, natural chemicals) can cause cancer. But as the NCI epidemiologists found on Long Island, trace exposures to "environmental toxins" generally can't be linked to cancer. Nevertheless, our built-in drive to identify patterns will guarantee that we will continue to seek someone to blame for our ills and that we will suffer through many more expensive, unnecessary, and self-defeating witch hunts for a long time to come.

Even the US National Cancer Institute denies that there is a "cancer epidemic." As the institute explains, the common misconception that we are experiencing a "cancer epidemic" stems largely from sensationalized media reports. "This only *appears* to be the case because the *number* of new cancer cases reported is rising as the population is both expanding and aging. Older people are more likely to develop cancer," notes the institute. "So as more and more members of a 75-million-strong 'baby-boomer' cohort begin shifting *en masse* to older, more cancer-prone ages, the *number* of new cancer cases is expected to increase in the next several decades."

If you are male in the United States your lifetime risk of developing cancer is approximately 1 in 2, and your risk of dying of cancer is 1 in 4. If you are female your lifetime risk of contracting cancer is 1 in 3, and your risk of dying of malignancy is 1 in 5. Is an especially toxic environment responsible for these grim statistics? Actually, no. What these statistics signal is that you are likely to live a long time. If you live long enough you will get cancer.

A look back in time is instructive. In 2012, the *New England Journal of Medicine* published a fascinating article comparing the annual death rates between 1900 and 2010. The annual death rate in 1900 from the top ten causes of death was 1,100 per 100,000 (the all-cause death rate was just over 1,700 per 100,000). Of those deaths, more than half were caused by infectious diseases. Pneumonia or influenza killed 202 per 100,000; tuberculosis 194 per 100,000; gastrointestinal infections 143 per 100,000; and diphtheria 40 per 100,000. What about cancer? Cancer accounted for just 64 deaths out of 100,000 in 1900.

By 2010, the top ten causes of death killed just over half as many Americans, at the rate of about 600 per 100,000 (the all-cause death rate was just shy of 800 per 100,000). In the list of contemporary leading causes of death, infectious diseases hardly figure at all in the *New England Journal of Medicine* statistics. In fact, the infectious diseases listed are pneumonia and influenza, which kill 16 per 100,000 annually now. Cancer? In 2010, the disease caused 186 out of 100,000 deaths annually, triple the number in 1900. That initially sounds terrible until one considers the fact that only 47 percent of Americans lived past age sixty in 1900. Today, 88 percent of Americans live past age sixty.

The median age at which cancer is diagnosed is sixty-five, and 53 percent of all cancers are diagnosed in people over age sixty-five. Seventy percent are diagnosed in people over age fifty-five. In 1929, the first year for which the US Centers of Disease Control has national data, average life expectancy in the United States was fifty-seven years.

Roughly speaking, this suggests that in the first decades of the twentieth century Americans were not living long enough for them to have developed around 75 percent of today's cases of cancer. That's why cancer was then comparatively rarer. And that's not taking into account the lower incidence rates that would likely have existed due to higher levels of physical activity and lower rates of obesity, cigarette smoking, and alcohol consumption that prevailed among Americans in the early part of the twentieth century. For example, the lung cancer death rate for men was just shy of 5 per 100,000 in 1930; by 1990, the rate had risen to 76 per 100,000.

Hormone Havoc: Half the Men Our Fathers Were?

While some environmental groups remain loyal to the old belief that exposures to trace amounts of synthetic chemicals are causing a cancer epidemic, others are now pushing the notion that these chemicals are generating hormone havoc. In *Silent Spring*, Carson does vaguely speculate that pesticide residues might increase endogenous estrogens, and she also worries about the exposures to synthetic estrogens in cosmetics, drugs, foods, and workplaces. She mentions in passing medical reports that suggest there is reduced sperm production among crop dusters. Nearly thirty years later, a group of Carson disciples meeting in 1991 at the Wingspread Center in Wisconsin under the auspices of the World Wildlife Fund developed their guru's suspicions about the effects of synthetic chemicals on sex hormones into the endocrine disruptor conjecture. The idea is that some synthetic chemicals harmfully produce the effects of estrogen, testosterone, and other hormones on human bodies and in wildlife.

I suspect the endocrine disrupting chemical (EDC) controversy will play out much the same way as the cancer controversy has. Basically it will turn out that some synthetic compounds in high doses will have deleterious effects on those exposed, but background exposures will have no detectable effects on the larger population.

As it happens, proponents of the idea that endocrine disruption is a major public health problem have issued a couple of consensus statements recently. For example, one such was issued in 2013 by a group working under the auspices of the United Nations Environment Programme. *The State of the Science of Endocrine Disrupting Chemicals* report begins by confidently asserting that many endocrine-related disorders are on the rise, including low semen quality and deformed penises, early breast development in girls, attention deficit and hyperactivity disorder in children, obesity and type 2 diabetes, and testicular, prostate, breast, and thyroid cancer. The "consensus" on all of these damaging effects was derived by combing selectively through the epidemiological literature. Interestingly, if you read through the data cited by the consensus architects, it becomes clear that the confident assertions in the *State of the Science* consensus are based on a lot of vigorous hand-waving and embarrassingly weak data.

For example, it bears noting that the *State of the Science* report itself admits that current research does not support the claim that synthetic endocrine disrupting chemicals are responsible for the trends that it identifies. The *State of the Science* report observes that "evidence linking estrogenic environmental chemicals with [breast cancer] is not available." In addition, the *State of the Science* report acknowledges that "there is very little epidemiological evidence to link EDC [endocrine disrupting chemical] exposure with adverse pregnancy outcomes, early onset of breast development, obesity or diabetes." Furthermore, the *State of the Science* report concedes, "There is almost no information about associations between EDC exposure and endometrial or ovarian cancer," and "no studies exist that explore the potential link between fetal exposure to EDCs and the risk of testicular cancer occurring 20–40 years later." Additionally, the *State of the Science* report notes that "high accidental exposures to PCBs [polychlorinated biphenyls] during fetal development or to dioxins in childhood increase the risk of reduced semen quality in adulthood." If PCBs do lower sperm quality, it's a problem of the past rather than the future, since the production of PCBs was banned in the United States in 1979. Let's look beyond the feeble findings of the *State of the Science* report and see what other research has to say about the asserted trends.

Penis Problems and the Sperm Apocalypse?

Alarm about allegedly falling sperm counts due to exposure to synthetic estrogens was first raised in a 1992 article by Scandinavian researchers, who reported there had been a decline of nearly 50 percent in fifty years. Ever ready to fan the flames of panic, the publicists at Greenpeace quickly initiated a clever campaign of advertisements declaring, "You're not half the man your father was."

Although the originators of the 1992 finding of falling semen quality and increasing penile anomalies continue to produce epidemiology to bolster those claims, many other researchers report contrary trends. For example, a 2013 comprehensive review of thirty-five sperm quality studies published after the purported decline was first announced in 1992 finds no such overall trend. The researchers report that eight studies involving a total of 18,109 men suggest a decline in semen quality; twenty-one studies encompassing 112,386 men show either no change or an increase in semen quality; and six studies involving 26,007 men show ambiguous or conflicting results. According to the researchers, the upshot is that "allegations for a worldwide decline in semen parameter values have not withstood scientific scrutiny."

What about the epidemic of deformed penises that endocrine disruptors are supposedly engendering? One of the more common birth defects in males is hypospadias, in which the urethral opening occurs elsewhere along the penis rather than at the tip. A 2012 comprehensive review of data on trends for this birth defect reported that "generalized statements that hypospadias is increasing are unsupported" and that "firm conclusions cannot be made regarding the association of endocrine-disrupting exposures with hypospadias." Another systematic review of studies looking for genetic and environmental influences on hypospadias found that "[w]hile genes involved in the aetiology of hypospadias have received a considerable amount of attention, research on environmental factors has been even more extensive. Despite the large number of studies, however, clear evidence for causal environmental factors is still lacking."

A similar 2009 report, "Rising Hypospadias Rates: Disproving a Myth," concluded, "A review of the epidemiologic data on this issue amassed to date clearly demonstrates that the bulk of evidence refutes claims for an increase in hypospadias rates."

Testicular cancer does appear to be increasing in many developed nations. Proponents of the endocrine disruption thesis have conducted epidemiological studies that weakly suggest that they do correlate with greater risk for testicular cancer. On the other hand, lots of other physiological phenomena also correlate with higher testicular cancer rates. A 2012 study reports that one of the stronger correlations uncovered by several studies is that the risk of testicular cancer increases with adult height. The researchers in that study find that men over six feet tall have a higher risk of testicular cancer, and they suggest that "the trend of increasing adult height and the increasing TC [testicular cancer] incidence are biologically interconnected with improved nutrition in early life." They note that while testicular cancer rates in Europe increased throughout the twentieth century, they stalled for the generation that suffered nutritional deprivation as a result of World War II. Interestingly, being overweight reduces the chances of testicular cancer, so one might think that as the obesity rate rises, the testicular cancer rate should fall.

The *State of the Science* consensus report notes, "The prevalence of obesity and type 2 diabetes has dramatically increased worldwide over the last 40 years." As noted above, the consensus report, however, admits there is very little epidemiological evidence linking endocrine disrupting chemicals to increasing obesity rates, but that does not stop researchers from trying to do so. A typical study, "Association of Endocrine Disruptors and Obesity: Perspectives from Epidemiologic Studies," published in 2010, found a correlation between excess weight and endocrine disrupting chemicals. However, that study begins by acknowledging that "changes in diet and physical activity are undoubtedly key causal factors related to the increase in obesity." Well, yes.

A far more plausible explanation for the rise in obesity can be found in a 2013 study that tracked the changes in the number of calories Americans consumed daily between 1971 and 2010. The researchers found that in 1971, Americans ate an average of 1,955 calories daily. That average rose to 2,269 per day by 2003 and has recently dropped a bit, to 2,195 calories daily. While Americans are eating more, they are also exercising less. A 2011 study reported, "Over the last 50 years in the U.S. we estimate that daily occupation-related energy expenditure has decreased by more than 100 calories, and this reduction in energy expenditure accounts for a significant portion of the increase in mean U.S. body weights for women and men." It is well established that being overweight escalates considerably the risk of type 2 diabetes. A 2012 workshop organized by the US National Institute of Environmental Health Sciences reviewed seventy-five studies relating exposures to synthetic endocrine disruptors to type 2 diabetes and concluded, "In no case was the body of data considered sufficient to establish causality." It's hard not to conclude that in comparison to eating more calories and increasingly sedentary lifestyles, the effects of endocrine disrupting chemicals on rising obesity and the prevalence of type 2 diabetes, if such effects exist, would be negligible.

The *State of the Science* report also asserts that there is a "trend towards earlier onset of breast development in young girls." Again, some epidemiologists have sought to identify synthetic endocrine disruptors as the possible villains behind this trend. However, considerable research persuasively suggests that earlier breast development and puberty onset in girls is linked to increases in body fat. In other words, rising obesity rates strongly correlate with earlier onset of puberty in girls. There does appear to be an epidemic of *diagnoses* of attention deficit hyperactivity disorder (ADHD), but many researchers question the claim that actual cases are increasing. A 2014 comprehensive review in the *International Journal of Epidemiology* analyzed 135 studies dealing with trends in ADHD and reports, "In the past three decades, there has been no evidence to suggest an increase in the number of children in the community who meet criteria for ADHD when standardized diagnostic procedures are followed."

A devastating article, "Endocrine Disruption: Fact or Urban Legend," by a team of European and American toxicologists, published in the journal *Toxicology Letters* in December 2013, reviews hundreds of studies on the effects of alleged endocrine disrupting chemicals. Their review of the science finds no effects on sperm counts, no effects on the rate of penile deformations, no effects on any cancer, no effects on diabetes, no effects on breast development, and . . . simply no effects whatsoever from exposures to chemicals, natural and synthetic, that emit a weak hormonal signal in very sensitive lab tests. Among many other things, the researchers point out that voluntary human exposures to powerful hormone modifying substances such as the estrogens in birth control pills have resulted in practically no significant health effects on either those who take them or their progeny. The effective potency of contraceptive pills is more than a million times greater than the potencies of the weak endocrine disrupting chemicals on which environmental activists focus. Taking this and much other evidence into account, the researchers assert that "the hypothesis that the negligible exposure of humans to chemicals of negligible hormonal potency could have an effect on human fertility is absurd, defying a scientific basis as well as common sense." They conclude that the man-made environmental disruptor hypothesis "has now been evaluated experimentally and epidemiologically for nearly 20 years and no convincing evidence has been found of an actual decline in human fertility, and even less of a causal relation with synthetic hormonally active substances."

The Problem of Epidemiology

Epidemiologists are hard at work sincerely trying to uncover relationships between all sorts of exposures and health outcomes. Many of the assertions made about the possible effects of endocrine disrupting chemicals in the *State of the Science* report are based on the results of observational studies by epidemiologists. The number of observational studies has increased from 80,000 in the 1990s to more than 260,000 in the first decade of the twenty-first century. Most observational studies are case-control studies in which epidemiologists identify a population that suffers from a disease or other condition and then attempt to match them with a similar population free of the disease or condition. They then look for differences in lifestyle, diet, or the environment that might account for the disease. This should work in theory, but the problems with controlling for biases in the data and for confounding factors are well known to epidemiologists. Confounding factors are variables that have been overlooked by researchers. Confounders can easily generate spurious associations.

For example, one famous study found a link between heavy coffee drinking and pancreatic cancer that disappeared once the smoking and alcohol consumption habits of coffee drinkers were taken into account. In addition, biases can creep in because it turns out that the control population differs in significant but unrecognized ways. For example, a finding that exposure to electromagnetic fields caused leukemia disappeared when differences in the incomes of the case population and the control population were taken into account. There is a well-known epidemiological relationship between poverty and cancer.

The problem is, epidemiologists generally find far more false positives than they do true positives—that is, they identify far more associations between phenomena than eventually are found to be the case. How do we know that there are far more false positives than true positives? Because the vast majority of epidemiological studies are not replicated. In other words, other, later researchers do not find that the risk factor identified in the initial observational study is in fact associated with a disease. S. Stanley Young of the US National Institute of Statistical Sciences estimates that only 5 to 10 percent of observational studies can be replicated.

In addition, there is a strong tendency among epidemiologists to publish only studies with positive results. Reporting only positive results (that is, a finding that some risk factor is associated with disease) skews the literature toward implying that various risk factors are more dangerous than is really the case. Others worry that researchers may try to please their sponsors, especially the government regulatory agencies that fund their research. As one anonymous researcher at the National Institute of Environmental Health Sciences told *Science,* "Investigators who find an effect get support, and investigators who don't find an effect don't get support. When times are rough it becomes extremely difficult for investigators to be objective."

Even with the best of scientific intentions, it is not easy to sort actual risk factors from the statistical background noise of confounders and researcher biases. "With epidemiology you can tell a little thing from a big thing. What's very hard to do is to tell a little thing from nothing at all," said Michael Thun, an American Cancer Society epidemiologist, in 1995. Former Boston University epidemiologist Samuel Shapiro agrees: "Epidemiologists have only primitive tools, which for small relative risks are too crude to enable us to distinguish between bias, confounding, and causation."

So, most epidemiologists will agree that one study that identifies a small effect means very little. However, if a number of studies consistently find a similar small relative risk for a factor, then perhaps the factor is causal. But consistency among studies can go only so far. If all of the studies have the same design, they could all be implementing the same biases and missing the same confounders and thus producing the same spurious positive results. That brings up the issue of pathological science.

Possibly Pathological Science?

In 1953, Nobel Prize–winning chemist Irving Langmuir identified what he called "pathological science," or "the science of things that aren't so." In his 1992 summary of Langmuir's insights, Denis Rousseau noted that "the first characteristic of pathological science is that the effect being studied is often at the limits of detectability or has very low statistical significance. Thus it can be difficult to do experiments that reliably test the effect." In such scientific situations "unconscious personal bias may affect the results."

Experimental researchers into the effects of endocrine disrupting chemicals have devised a set of exquisitely sensitive cell-based and animal model tests to detect their subtle influences. They then turn around and justify the relevance of the barely detectable responses conjured from their hypersensitive assays by citing supposed endocrine disruptor effects found in human epidemiological studies. Amusingly, some epidemiologists return the favor and suggest that the plausibility of their very weak findings is justified by the experiments using the sensitive assays. Neither bothers to explain why the results of these highly sensitive tests have any relevance to human health.

Rousseau adds, "Because the effect is weak or of such low statistical significance, there may be no consistent relationship between the magnitude of the effect and the causative agent. Increasing the strength of the causative agent may not increase the size of the effect." Another hallmark of pathological science is that its effects can be elicited by only some researchers, who are then unable to communicate how they achieve them to other researchers. Proponents of endocrine disruption generally respond that outside researchers who can't replicate their findings are simply not careful enough. Naturally, practitioners of pathological science are impervious to critiques and accusations of sloppiness on their part from other scientists.

In addition, practitioners of pathological science propose eccentric theories, positing mechanisms that appear nowhere else in related sciences. In this case, champions of the environmental endocrine disruption hypothesis reject the standard thinking in toxicology, according to which the biological effects of a substance increase as the dose increases, often summarized as "the dose makes the poison." Instead, they propose the concept of nonmonotonic dose response, in which exposures to allegedly endocrine disrupting compounds exhibit a U-shaped dose response curve—that is to say, barely detectable doses produce a big effect that falls off as the dose increases to a certain level, at which point bigger doses also produce big effects.

Perhaps environmental endocrine disruptor researchers have uncovered a real phenomenon, but most toxicologists doubt it. "Although this hypothesis is consistent with the ideas of homoeopathy, it contradicts centuries of toxicological and pharmacological experience demonstrating that active substances produce a specific dose-response in the affected organism," tartly asserted one group of critical toxicologists. Homeopathy is a medical pseudoscience in which the alleged remedies are so diluted that they often do not contain a single molecule of any supposed therapeutic substance. Low-dose effects indeed.

To explain how researchers and whole fields of science can end up studying phenomena that don't actually exist, Stanford University biostatistician John Ioannidis fancifully describes the highly active areas of scientific investigation on Planet F345 in the Andromeda Galaxy. The Andromedean researchers are hard at work on such null fields of study as "nutribogus epidemiology, pompompomics, social psychojunkology, and all the multifarious disciplines of brown cockroach research—brown cockroaches are considered to provide adequate models that can be readily extended to humanoids."

The problem is that the Andromedean scientists don't know that their data dredging and highly sensitive nonreplicated tests are massively producing false positives. In fact, the Andromedean researchers have every incentive—publication pressure, tenure, and funding—to find effects, the more extravagant the better. But in fact, the manufactured discoveries are just estimating the net bias operating in each of these "null fields."

During the past twenty years hundreds of millions of euros and dollars of taxpayer money have been spent on endocrine disruptor research with essentially no results. In a remarkable and thorough 2013 scientific review article, a team of toxicologists bluntly suggests that all this funding has likely produced "a vested interest of scientists in the endocrine disruption field to keep the endocrine disruption hypothesis on the agenda in order to stay in business." Decades of research and hundreds of millions of dollars in funding have resulted in the publication of more than 4,000 different articles. "Taking into account the large resources spent on this topic, one should expect that, in the meantime, some endocrine disruptors that cause actual human injury or disease should have been identified," the researchers argue. "However, this is not the case. To date, with the exception of natural or synthetic hormones, not a single, man-made chemical endocrine disruptor has been identified that poses an identifiable, measurable risk to human health." They damningly add, "Certainly, there has been much media hype about imaginary health risks from bisphenol A, parabens, or phthalates. However, no actual evidence of adverse human health effects from these substances has ever been established. To the contrary, there is increasing evidence that their health risks are absent or negligible—or imaginary."

As Denis Rousseau cogently reminds us, his description of "science gone bad is not a portrait of deliberately fraudulent behavior. Pathological science arises from self-delusion—cases in which scientists believe that they are acting in a methodical, scientific manner but instead have lost their objectivity. The practitioners of pathological science believe that their findings simply cannot be wrong. But any ideas can be wrong and any observation can be misinterpreted."

On the basis of the evidence so far, there is a very good chance that the study of endocrine disruption will ultimately turn out to be what Ioannidis calls a null field. In which case, apocalyptic researchers will have provoked the public and policymakers into spending a great deal of time, energy, funding, and regulatory attention on another exaggerated environmental scare.

5

The Attack of the Killer Tomatoes?

THE CRYSTALS AND GEMS GALLERY IN HANALEI, a trendy little town on Kauai, displayed several posters protesting GMOs and offered flyers urging a ban on biotech crops. The gallery is the sort of place where, when my wife picked up an attractive stone and asked a clerk what it was, the reply came back, "Do you mean, what does it *do*?" Apparently, that particular rock can dispel negativity.

After being advised on the therapeutic properties of various crystals, we asked the clerk what all the anti-biotech literature around the shop was about. Among other things, she informed us that biotech crops cause cancer, stating emphatically that Kauai's cancer rates were exceptionally high, especially among people who live close to the seed company fields on the island where biotech crop varieties are grown.

As it happens, the state Health Department reported earlier in 2013 that "overall cancer incidence rates (all cancers combined) were significantly lower on Kauai compared to the entire state of Hawaii." Nor did the department find higher rates of cancer in those districts where the seed company farms are located.

Some readers will recognize that the title of this chapter is taken from the 1977 comedy horror movie of the same name in which giant mutant tomatoes nearly destroy humanity. Unfortunately, it is not just winsome clerks in crystal shops who fear that modern biotech crops are the moral equivalent of homicidal tomatoes. Major environmental lobbying groups claim to be troubled by them too. For example, Doug Gurian-Sherman, a scientist formerly with the Union of Concerned Scientists and now at the Center for Food Safety, asserted in 2014, "There's no real consensus on GMO crop safety." In 2012, a statement issued by the Friends of the Earth in Europe demanded a moratorium on all foods derived from biotech crops. That FOE statement declared, "As well as posing unnecessary risks to human health, Friends of the Earth Europe believes GMOs destroy biodiversity, lead to increased costs for conventional farmers, increase corporate control of the food chain, and fail to combat global hunger." In 2013, Daniel M. Ocampo, Sustainable Agriculture and Genetic Engineering Campaigner for Greenpeace in Southeast Asia, stated, "There is no scientific proof that GMOs pose no danger to human health and the environment." Ocampo added, "Even the scientific community is divided on whether GMOs are safe."

All of these statements by representatives from the world's leading environmentalist organizations are false. There is, in fact, a broad scientific consensus that modern biotech crop varieties are safe for people and the environment. Why, then, are prominent environmentalist groups opposed to this technology? That takes a bit of history to explain.

Biotech Born Precautionary

Biotechnology was born precautionary, and that's been a problem ever since. Modern biotechnology got its start in 1971 when Stanford University biochemist Paul Berg figured out how to splice segments of

DNA together. He called this process *recombining*, and it enabled researchers to move genes from one organism to another. In 1980 Berg won the Nobel Prize in chemistry for this discovery. This new capability made some researchers uncomfortable, so a committee of prominent molecular biologists published a letter in the journal *Science* in July 1974 asking for a worldwide moratorium on certain types of gene-splicing experiments. The moratorium was supposed to last until the hazards that might be posed by gene splicing could be assessed. The researchers also asked that the National Institutes of Health devise a set of safety procedures for working with recombinant DNA. This was the first self-imposed ban on basic research in the history of science and it lasted two years.

Naturally, the moratorium attracted the attention of a wide variety of activists, many of whom worried about researchers playing God with nature. Some warned that super-plagues would escape the laboratories. Alarmed members of Congress including Senator Ted Kennedy (D-MA) and Representative Al Gore (D-TN) proposed significant and burdensome regulation, including the creation of a National Biohazards Commission modeled on the Nuclear Regulatory Commission.

In 1976, *The New York Times Magazine* published an alarming front-page article, "New Strains of Life—or Death," by Cornell University biochemist Liebe Cavalieri. Cavalieri asserted, among other horrors, that gene splicing could lead to accidental outbreaks of infectious cancer. "In the case of recombinant DNA, it is an all or none situation—only one accident is needed to endanger the future of mankind," he warned.

Also in 1976, Alfred Vellucci, the mayor of Cambridge, Massachusetts, guided by the left-leaning group Science for the People, wanted to ban gene-splicing research in his city. Of course, Cambridge is home to Harvard University and the Massachusetts Institute of Technology. "We want to be damned sure the people of Cambridge won't be affected by anything that could crawl out of that laboratory," Vellucci told *The New York Times*. He added, "They may come up with a disease that can't be cured—even a monster. Is this the answer to Dr. Frankenstein's dream?" There is no little irony that today Cambridge promotes itself as "one of the world's major biotech centers." Needless to say, more than forty years after gene splicing was invented, no plagues, much less epidemics of infectious cancer, have emerged from the world's biotech labs.

In the context of this furor, some 140 molecular biologists convened in 1975 at the Asilomar Conference Grounds in Pacific Grove, California, to draft guidelines for conducting gene-splicing experiments. They self-consciously thought that they were avoiding what they saw as the mistakes made a generation earlier by Manhattan Project nuclear physicists when they unleashed the power of the atom. The initially restrictive guidelines have been greatly relaxed, not least because it turns out that microorganisms are natural and promiscuous exchangers of genes.

Reflecting later on the hysteria and rush to regulate, James Watson, codiscoverer of the double-helix structure of DNA, for which he won the Nobel Prize, succinctly noted, "Scientifically I was a nut. There is no evidence at all that recombinant DNA poses the slightest danger." Similarly, biophysicist Burke Zimmerman, who participated in the congressional debates over regulating biotechnology, concluded, "In looking back, it would be hard to insist that a law was necessary, or, perhaps, that guidelines were necessary."

"We Shall Not Be Cloned"

However, once fears are raised, they are hard to allay, especially if some groups find them useful for advancing other agendas. One master promoter of fear is onetime radical organizer Jeremy Rifkin, who early on became an anti-biotech campaigner. In 1977, Rifkin led a group of protesters into a meeting of molecular biologists at the National Academy of Sciences, where they joined hands singing, "We shall not be cloned." Also in 1977, Rifkin and his fellow activist Ted Howard published *Who Should Play God? The Artificial Creation of Life and*

What It Means for the Future of the Human Race. "The traditional notion of ruthlessly exploiting and controlling nature in the name of progress is being challenged by an environmentalist creed that emphasizes a reintegration into the ecosystem," wrote Rifkin and Howard. They also railed against "unbridled scientific and technological progress" and creeping "corporate hegemony" and called for a "new spiritual awakening," which would produce "a fundamental change in the values and institutional relationships of American society."

In 1984, Rifkin penned a semi-mystical tract, *Algeny: A New Word—A New World*, in which he disparaged biotechnology by likening it to medieval alchemy. Modern technology was alienating humanity from nature, argued Rifkin. "Humanity seeks the elation that goes with the drive for mastery over the world," he asserted. "Nature offers us the sublime resignation that goes with an undifferentiated participation in the world around us."[10] Nature may indeed offer sublimity, but it also deals out plague, floods, droughts, and starvation quite liberally. In a review, Harvard University paleontologist Stephen Jay Gould correctly decried *Algeny* "as a cleverly constructed tract of anti-intellectual propaganda masquerading as scholarship."

Rifkin established the Foundation on Economic Trends, from which he launched many protests and legal challenges against the nascent biotechnology industry. For example, he challenged researchers who wanted to field-test a genetically modified version of the ubiquitous *Pseudomonas syringae* bacteria as a way to prevent frost damage to crops. The natural bacteria carry a gene that causes ice crystals to form around them when temperatures drop below freezing so that they can dine on the frost-damaged plants. Researchers had simply removed the gene and wanted to spread the modified bacteria over crops in the field to see if it would protect against frost damage. In what would later become a standard operating procedure for opponents of biotech, members of the group Earth First! ripped up one of the test plots in 1987. Eventually research showed that the modified bacteria did protect against frost damage, but efforts to commercialize it were dropped as threats to vandalize test plots persisted.

Rifkin was also a big player in the campaign that delayed the Food and Drug Administration's approval of a biotech version of the hormone bovine somatotropin (BST) to boost milk production by as much as 25 percent. In 1986 the FDA had determined that BST was safe, since it is inactive in people. In addition, tests cannot differentiate milk from dairy cows treated with the hormone from milk from cows not so treated. The FDA finally approved BST in 1993 after finding that "milk from treated cows was safe for human food." Every subsequent review by the agency has confirmed its safety. Spooked by the controversy and eager to avoid producing surpluses of highly subsidized meat and milk, regulators in Europe took the precautionary step of banning its use and the import of meat from treated cows. In 1999, the World Trade Organization ruled that the European ban on imports was based on phony safety concerns, the chief goal of which was to shield its farmers from competition and to protect its system of bloated farm subsidies.

In 1994, Calgene's delayed-ripening Flavr Savr tomato became the first genetically modified food crop grown and consumed in an industrialized country. Unfortunately, the tomato did not actually taste all that good, so the product failed among consumers. The first commercially successful modern biotech crops were planted in the United States in 1996. They included corn, potato, and cotton varieties that had been enhanced to resist pests and soybeans with an added herbicide resistance trait.

Modern Crop Biotechnology Arrives

For pest resistance, researchers installed versions of a gene from the *Bacillus thuringiensis* (Bt) that kills insect pests with alkaline digestive systems. *Bacillus thuringiensis* has been used for decades in agriculture, especially by organic farmers, as an insecticide. Pest insects, usually caterpillars, ingest the bacteria, and their highly alkaline guts activate it to release protein crystals that poke holes in their digestive tracts, causing them to die. Studies accepted by the US Environmental Protection Agency (EPA) show that Bt protein is safe for people to eat; it is degraded by human gastric fluids within thirty seconds. The herbicide resistance trait is derived from a gene isolated from another soil bacterium that prevents the glyphosate herbicide (Roundup) from interfering with the production of plant-specific amino acids needed for growth and development. The EPA pointed out that the bacterial gene that confers herbicide resistance differs little from the same gene that all plants already have. Reviewing the test data, the agency also found that the herbicide resistance trait is not toxic to mammals or birds.

Since 1996, biotech crops have been adopted faster than any previous agricultural technology. In 1996, farmers planted 4.2 million acres of biotech crops. In 2013, 18 million farmers in twenty-seven countries planted 433 million acres, a hundredfold increase over eighteen years. Some 16.5 million of the farmers growing biotech crops are resource-poor farmers living in developing countries. Between 1996 and 2012, economic gains to farmers amounted to nearly $120 billion, 58 percent from saved inputs such as less pesticide, less plowing, and less labor. The remainder came from higher yields, totaling about 377 million tons over the past eighteen years. Due to these boosted yields, farmers did not have to plow up to an additional 300 million acres to keep up with the growing demand for food. This helped to preserve tropical forests and other biodiversity-rich ecosystems. Furthermore, by switching to biotech crops, farmers were able to cut their use of pesticides over the period by more than 1 million pounds of active ingredient, a reduction of nearly 9 percent.

The Politics of Anti-Biotech Environmentalism

Given these well-established advantages, including benefits to the natural environment such as lessening pesticide use, preventing soil erosion, and saving wild lands from being converted to farming, why do many leading environmentalist groups so strenuously oppose bio-tech crops? The chief problem is that just as biotech crops were being commercialized in the 1990s, they ran into a perfect storm of food and health safety scandals in Europe. In order to protect beef farmers and suppliers, British food safety authorities were downplaying the dangers of "mad cow" disease, which was spread by feeding cattle infected sheeps' brains. Even though there were tests that could have identified donations from people infected with HIV, the agency in charge of French public health was permitting transfusions of contaminated blood. And the asbestos industry had exercised undue influence over its regulators in evaluating the risks posed by that mineral.

These feckless incidents "led to strong distrust and caused people to think that firms and public authorities sometimes disregard certain health risks in order to protect certain economic or political interests," argues French National Institute of Agricultural Research analyst Sylvie Bonny. Consequently, Bonny notes that these events "increased the public's attention to critical voices, and so the principle of precaution became an omnipresent reference." So when biotech crops were being introduced, much of the European public was primed to credit any claims that this new technology might carry hidden risks.

Bonny points out that opposition to biotech crops arose first among "ecologist associations," including Greenpeace and Friends of the Earth. In fact, hyping their opposition to biotech crops served as a lifeline to organizations like Greenpeace. Bonny notes that in the late 1990s, Greenpeace in France was experiencing a serious falloff in membership and donations, but the GMO issue rescued the group. "Its anti-GMO action was instrumental in strengthening Greenpeace-France which had been in serious financial straits," she reports. It should always be borne in mind that environmentalist organizations raise money to support themselves by scaring people. More generally, Bonny observes, "For some people, especially many activists, biotechnology also symbolizes the negative aspects of globalization and economic liberalism." She adds, "Since the collapse of the communist ideal has made direct opposition to capitalism more difficult today, it seems to have found new forms of expression including, in particular, criticism of globalization, certain aspects of consumption, technical developments, etc."

These concerns are obviously well beyond any scientific considerations regarding the safety of biotech crops for health and the environment. In any case, as the result of the institutional survival imperatives of environmentalist and social movement organizations, opposition to biotech crops became a major strand of their anti-globalization ideology. That had significant negative consequences for the acceptance of this useful technology around the world, especially in poor countries. To illustrate the destructive effects of this ideology, here are a few disturbing vignettes.

In October 1999, a cyclone killed 10,000 people and left 10 million homeless when it slammed into India's eastern coastal state of Orissa (now known as Odisha). In the aftermath, nonprofit private aid agencies distributed a high-nutrition mixture of corn and soy meal to the hungry in Orissa. This act of charity outraged Indian ecofeminist Vandana Shiva. "We call on the government of India and the state government of Orissa to immediately withdraw the corn-soya blend from distribution," demanded Shiva, director of the New Delhi–based Research Foundation for Science, Technology, and Ecology. "The U.S. has been using the Orissa victims as guinea pigs for GM [genetically modified] products which have been rejected by consumers in the North, especially Europe."

In response to Shiva's unscientific rant, Per Pinstrup-Andersen, then director general of the International Food Policy Research Institute, observed: "To accuse the US of sending genetically modified food to Orissa in order to use the people there as guinea pigs is not only wrong; it is stupid. Worse than rhetoric, it's false. After all, the US doesn't need to use Indians as guinea pigs, since millions of Americans have been eating genetically modified food for years now with no ill effects."

In 2002, 2.5 million Zambians were on the edge of starvation. To alleviate the impending famine, the United States sent food aid, including tons of biotech corn. The president of Zambia refused the food. "We would rather starve than get something toxic," said Zambian President Levy Mwanawasa. Mwanawasa also stated, "Simply because my people are hungry, that is no justification to give them poison, to give them food that is intrinsically dangerous to their health." On the other hand, as the *Los Angeles Times* reported, some of his citizens disagreed. "I would rather eat that maize than die because the government has no alternative to the hunger problem," said Bweengwa Nzala, a twenty-eight-year-old farmhand. The reason that Mwanawasa thought biotech crops were toxic had everything to do with the propaganda peddled by international environmental organizations and the European Union's official efforts to get developing nations to adopt anti-biotech regulations. After all, Americans had been safely eating foods made with ingredients from biotech corn for seven years at that point.

Also in 2002 the United Nations convened a World Summit on Sustainable Development in Johannesburg, South Africa. As is usual with such confabs, the UN arranged for youth representatives to meet. At one such session, a young Kenyan announced his opposition to GMOs. Why? Because he had heard that GMOs weakened the immune systems of Africans so that they would more easily succumb to AIDS. From whom could he have gotten this disinformation? What could be more demoralizing than to believe that white scientists from rich countries had devised a technology that aimed to kill you?

In 2003, I watched a Brazilian member of the Friends of the Earth repeatedly yell at a group of poor Mexican women to whom food made with biotech ingredients was being distributed that it was "contaminated" and "toxic" and would harm their children. The food included boxes of Kellogg's Corn Flakes. The good news is that the women ignored him and his fellow activists and took the food.

As the opening quotations from representatives of leading environmental lobbying groups show, they still are fighting crop biotechnology. Let us go through many of the claims made by the activists and see what the science shows. First, what about the activist assertion that there is "no real consensus on GMO crop safety"? Flatly false.

The Overwhelming Scientific Consensus
on Biotech Crop Safety

No one has ever gotten so much as a cough, sneeze, sniffle, or stomachache from eating foods made with ingredients from modern biotech crops. Every independent scientific body that has ever evaluated the safety of biotech crops has found them to be safe for humans to eat.

"We have reviewed the scientific literature on GE [genetically engineered] crop safety for the last 10 years that catches the scientific consensus matured since GE plants became widely cultivated worldwide, and we can conclude that the scientific research conducted so far has not detected any significant hazard directly connected with the use of GM crops," asserted a team of Italian university researchers in September 2013. And they should know, since they conducted the largest ever survey of scientific studies—more than 1,700—that evaluated the safety of biotech crops.

A statement issued by the board of directors of the American Association for the Advancement of Science, the largest scientific organization in the United States, on October 20, 2012, point-blank asserted that "contrary to popular misconceptions, GM crops are the most extensively tested crops ever added to our food supply. There are occasional claims that feeding GM foods to animals causes aberrations ranging from digestive disorders, to sterility, tumors and premature death. Although such claims are often sensationalized and receive a great deal of media attention, none have stood up to rigorous scientific scrutiny." The AAAS board concluded, "Indeed, the science is quite clear: crop improvement by the modern molecular techniques of biotechnology is safe."

In July 2012, the European Commission's chief scientific adviser, Anne Glover, declared, "There is no substantiated case of any adverse impact on human health, animal health, or environmental health, so that's pretty robust evidence, and I would be confident in saying that there is no more risk in eating GMO food than eating conventionally farmed food." At its annual meeting in June 2012, the American

Medical Association endorsed a report arguing against the labeling of bioengineered foods from its Council on Science and Public Health. The report concluded, "Bioengineered foods have been consumed for close to 20 years, and during that time, no overt consequences on human health have been reported and/or substantiated in the peer-reviewed literature." In December 2010, a European Commission review of 130 EU-funded biotechnology research projects, covering a period of more than 25 years and involving more than 500 independent research groups, found "no scientific evidence associating GMOs with higher risks for the environment or for food and feed safety than conventional plants and organisms."

A 2004 report from the National Academy of Sciences (NAS) concluded that "no adverse health effects attributed to genetic engineering have been documented in the human population." In 2003 the International Council for Science, representing 111 national academies of science and 29 scientific unions, found "no evidence of any ill effects from the consumption of foods containing genetically modified ingredients." The World Health Organization states, "No effects on human health have been shown as a result of the consumption of such foods by the general population in the countries where they have been approved."

A 2002 position paper by the Society of Toxicology found that "[t]he level of safety of current BD [biotechnology-derived] foods to consumers appears to be equivalent to that of traditional foods." In 2002, the US Government Accountability Office (GAO) reviewed the scientific literature and sought expert advice about the safety of genetically modified foods. The GAO concluded, "Biotechnology experts believe that the current regimen of tests has been adequate for ensuring that GM (genetically modified) foods marketed to consumers are as safe as conventional foods." The experts with whom the GAO consulted also pointed out that "there is no scientific evidence that GM foods cause long-term harm, such as increased cancer rates," and that "there is no plausible hypothesis of harm." GM foods might have adverse effects if they produced harmful proteins that remained stable during digestion.

However, the GAO noted that the proteins produced through genetic enhancement are in fact rapidly digested.

In 2000 the report *Transgenic Plants and World Agriculture*, issued under the auspices of seven national academies of science, including the US National Academy of Sciences and the British Royal Society, found that "no human health problems associated specifically with the ingestion of transgenic crops or their products have been identified." In a 2014 op-ed in *The New York Times*, even the notorious food puritan Mark Bittman declared of crop biotechnology that "the technology itself is not even a little bit nervous making," adding, *"the technology itself has not been found to be harmful* [emphasis his]."

Fringe Anti-Biotech Science

Unfortunately, there is no shortage of fringe scientists to gin up bogus studies suggesting that biotech crops are not safe. My personal favorite in this genre is Russian researcher Irina Ermakova's claim, unpublished in any peer-reviewed scientific journal, that eating biotech soybeans turned mouse testicles blue.

One widely publicized specious study was done by the French researcher Gilles-Éric Séralini and his colleagues. They reported that rats fed herbicide resistant corn died of mammary tumors and liver diseases. Séralini is the president of the scientific council of the Committee for Research and Independent Information on Genetic Engineering, *which describes itself as an* "independent non-profit organization of scientific counter-expertise to study GMOs, pesticides and impacts of pollutants on health and environment, and to develop non polluting alternatives." The committee clearly knows in advance what its researchers will find with regard to the health risks of biotech crops. But when truly independent groups, such as the European Society of Toxicologic Pathology and the French Society of Toxicologic Pathology, reviewed Séralini's study, they found it to be meretricious rubbish. Six French academies of science issued a statement declaring that the journal should never have published such a low-quality study and ex- coriating Séralini for orchestrating a media campaign in advance of publication. The European Food Safety Authority's review of the Seralini study "found [it] to be inadequately designed, analysed and reported."

In November 2013, *Food and Chemical Toxicology*, the journal that published the Séralini research, decided to retract the badly flawed study. All to the good. Sadly, however, such junk science has real-world consequences, since Seralini's article was apparently cited when Kenya made the decision to ban the importation of foods made with biotech crops. Unhappily, a different low-quality journal, which exercised no additional peer review, decided to republish Séralini's paper in June 2014.

Climate Consensus Yes—Biotech Consensus No

Leading environmentalist organizations insist that the public and policymakers accept the strong consensus regarding the reality of man-made climate change. "You have the strongest consensus we have seen in the science community about global climate change since the conclusion that tobacco caused lung cancer," asserts Union of Concerned Scientists (UCS) president Kevin Knobloch. Greenpeace also argues, "There is, in fact, a broad and overwhelming scientific consensus that climate change is occurring, is caused in large part by human activities." And Friends of the Earth has gone after ExxonMobil because it "has repeatedly attempted to undermine the scientific consensus on climate change and actively resisted attempts to limit carbon dioxide emissions through law." Good for them for defending the research results of climate science. But for these same environmentalist groups, not all scientific consensuses are equal.

Oddly, in trying to rebut the established scientific consensus with regard to the safety of biotech crops, the Center for Food Safety analyst Gurian-Sherman linked to a "statement" issued in October 2013 by the European Network of Scientists for Social and Environmental Responsibility asserting that no such consensus exists. Who is ENS-SER? A collection of longtime foes of agricultural biotechnology. The disingenuous statement has so far been signed by fewer than three hundred scientists, including such anti-biotech luminaries as Charles Benbrook, Vandana Shiva, Gilles-Éric Séralini, and Gurian-Sherman himself. Referring to this declaration as evidence against biotech crop safety is akin to citing a statement from tobacco company scientists asserting that cigarette smoking isn't a risk factor for lung cancer.

Environmental activists regularly accuse those who question the climate change consensus of bad faith and worse. But aren't they exhibiting a similar bad faith when they reject the broad scientific consensus on genetically modified crops? It is just pathetic that major environmentalist organizations reject mainstream science with regard to biotech crop safety, and further, it is a colossal betrayal of the trust of their members.

Biotech Crops and Pesticide Use

Another oft-heard claim by activists is that GMOs increase herbicide use. First, so what? This claim is simply an attempt to mislead people into thinking that more herbicide use must somehow be more dangerous. As a US Department of Agriculture report has noted, planting herbicide-resistant biotech crops enables farmers to substitute the more environmentally benign herbicide glyphosate (commercially sold as Roundup) for "other synthetic herbicides that are at least 3 times as toxic and that persist in the environment nearly twice as long as glyphosate." Glyphosate has very low toxicity, breaks down quickly in the environment, and enables farmers to practice conservation tillage, which reduces topsoil erosion by up to 90 percent. So the net environmental effect is quite positive.

However, it must be admitted that there are few honest brokers when it comes to this issue. Most of the research on biotech crops and herbicides is underwritten by either activist groups or industry. I have drawn my own conclusions, but I provide a fairly comprehensive review of the various studies on this question below.

When it comes to biotech crops and pesticide use data, the go-to guy for anti-biotech activists is Charles Benbrook. After a long career with various anti-biotech groups, Benbrook now serves as a research professor in the Center for Sustaining Agriculture and Natural Resources at Washington State University. He has an extensive history of publishing studies allegedly showing that the adoption of biotech crops boosts the use of pesticides. Four years after commercial biotech crops were first planted in the United States, for example, he concluded in 2001 that herbicide use had "modestly increased." Benbrook's article contradicted research published the year before by scientists with the US Department of Agriculture, who had found that biotech crops had reduced pesticide applications.

In a 2004 report funded by the Union of Concerned Scientists, Benbrook asserted that "GE [genetically engineered] corn, soybeans, and cotton have led to a 122 million pound increase in pesticide use since 1996." In contrast, a 2005 study in *Pest Management Science,* by a researcher associated with the pesticide lobby group CropLife, reported that planting biotech crops had "reduced herbicide use by 37.5 million lbs." A 2006 study done for the self-described nonadvocacy think tank National Center for Food and Agricultural Policy, founded in 1984 with a grant from the W. K. Kellogg Foundation, reported that planting biotech crops in the United States had in 2005 reduced herbicide use by 64 million pounds and insecticide applications by about 4 million pounds. Another 2007 study, by a team of international academic researchers led by Gijs Kleter from the Institute of Food Safety at Wageningen University in the Netherlands, concluded that in the United States, crops genetically improved to resist herbicides used 25 to 30 percent less herbicide per hectare than conventional crops did. In 2009, Benbrook issued a report for the anti-GMO Organic Center claiming that "GE crops have been responsible for an increase of 383 million pounds of herbicide use in the U.S. over the first 13 years of commercial use of GE crops."

Benbrook's latest study, issued in 2012, found that the adoption of pest-resistant crops had reduced the application of insecticides by 123 million pounds since 1996 but increased the application of herbicides by 527 million pounds, an overall increase of about 404 million pounds of pesticides. The media—including *Mother Jones*'s ever-credulous anti-biotech advocate Tom Philpott—reported these results unskeptically.

Benbrook largely got his 2012 results by making some strategic extrapolations of herbicide use trends to make up for missing data from the US Department of Agriculture.

Meanwhile, a 2012 study by Graham Brookes and Peter Barfoot at the PG Economics consultancy found planting modern biotech crop varieties had globally cut pesticide spraying by 997 million pounds from 1996 to 2010, an overall reduction of 9.1 percent. Brookes and Barfoot calculated the amount of pesticide used by multiplying the acreage planted for each variety by the average amounts applied per acre.

In May 2014 the USDA's National Agricultural Statistics Service issued its comprehensive *Pesticide Use in U.S. Agriculture: 21 Selected Crops, 1960–2008* report updating national herbicide and insecticide usage trends. The USDA finds that herbicide usage peaked at 478 million pounds in 1981—a decade and half prior to the introduction of the first biotech crop varieties—and fell to 394 million pounds in 2008. By the way, farmers applied 409 million pounds of herbicides to their crops in 1996. So instead of a massive increase in herbicide spraying, as claimed by Benbrook, the USDA actually reports a modest decline. Insecticide applications peaked in 1972 at 158 million pounds, dropping to 29 million pounds in 2008. It's worth noting that the insecticide DDT accounted for 11 percent of all agricultural pesticides used in 1972. Since biotech crops can protect themselves against insect pests, there is far less need for farmers to spray their crops.

This conclusion was further bolstered in a November 2014 study by German university researchers that reviewed 147 agronomic studies of pesticide use trends in biotech crops. They reported that genetic modification (GM) technology has increased crop yields by 21 percent, largely by lowering crop damage from pests. In addition, biotech crops "have reduced pesticide quantity by 37 percent and pesticide cost by 39 percent." While noting that biotech seeds are more expensive than those of conventional varieties, those costs are more than compensated for through savings in chemical and mechanical pest control. Consequently, they found that "average profit gains for GM-adopting farmers are 69 percent." More yield and lower pesticide applications means less potential damage to the natural environment. And more profits for farmers too! What's not to like?

Biotech Crop "Side Effects"

The anti-biotech Institute for Responsible Technology claims, "By mixing genes from totally unrelated species, genetic engineering unleashes a host of unpredictable side effects." Not really.

All types of plant breeding—conventional, mutagenic, and biotech— can, on rare occasions, produce crops with unintended consequences. The 2004 NAS report cited earlier includes a section comparing the unintended consequences of each approach; it concludes that biotech is "not inherently hazardous." Conventional breeding transfers thousands of unknown genes with unknown functions along with desired genes, and mutation breeding induces thousands of random mutations via chemicals or radiation. In contrast, the NAS report notes that biotech is arguably "more precise than conventional breeding methods because only known and precisely characterized genes are transferred."

The case of mutation breeding is particularly interesting. In that method, researchers basically blast crop seeds with gamma radiation or bathe them in harsh chemicals to produce thousands of uncharacterized mutations, then plant them to see what comes up. The most interesting new mutants are then crossed with commercial varieties, which are then released to farmers. The Food and Agriculture Organization's Mutant Varieties Database offers more 3,000 different mutated crop varieties to farmers. Many of these mutated varieties are planted as organic crops. Among the more recent new mutant offerings are two corn varieties, Kneja 546 and Kneja 627. Whatever genetic changes have been wrought in these corn varieties by induced mutagenesis, they must be far less known to researchers than any changes made to standard-issue biotech crops, yet these mutants get practically no regulatory scrutiny or activist censure.

The contention here is not that mutation breeding is inherently dangerous. Given its solid record of eighty years of safety, it's not. The point is that the more precise methods of modern gene splicing are even safer than that.

The Institute for Responsible Technology warns that producing biotech crops can produce "new toxins, allergens, carcinogens, and nutritional deficiencies." There is no evidence for any of this. Consider the panic back in 2000 over StarLink corn, in which a biotech variety approved by the EPA as feed corn got into two brands of taco shells. Some twenty-eight people claimed that they had experienced allergic reactions to eating "contaminated" tacos. The Centers for Disease Control and Prevention tested their blood and found that none reacted in a way that suggested an allergic response to StarLink.

The Union of Concerned Scientists cites a 1996 experiment in which researchers added a Brazil nut protein to a soybean variety that did produce allergic reactions. The company naturally did not proceed and that variety never made it out of the laboratory. One would think that that would actually be an example of the care with which biotech crop researchers test their products to make sure that they are safe before they are commercialized.

As far as cancer goes, it is worth noting that even as Americans have chowed down on billions of biotech meals, the age-adjusted cancer incidence rate has been going down (see Chapter 4). In fact, research shows that biotech corn engineered to resist insects is much lower in potent cancer-causing mycotoxins.

Biotech Crops and the Environment

The Institute for Responsible Technology recycles the fable that biotech crops harm monarch butterflies. This particular meme had its origins in 1999 when a researcher at Cornell University poisoned monarch butterfly caterpillars in his laboratory by forcing them to eat milkweed leaves coated with pollen from an insect-resistant corn variety. Of course the larvae died, since the *Bacillus thuringiensis* gene inserted into the corn specifically targets caterpillar pests like rootworms.

Countering misinformation takes a lot of work, but eventually the *Proceedings of the National Academy of Sciences* published a series of articles evaluating the effects of biotech corn on monarch butterflies in the wild. The researchers described the product's impact on monarch butterfly populations as "negligible." A 2011 review of more than 150 scientific articles found that "commercialized GM crops have reduced the impacts of agriculture on biodiversity, through enhanced adoption of conservation tillage practices, reduction of insecticide use and use of more environmentally benign herbicides, and increasing yields to alleviate pressure to convert additional land into agricultural use."

Consider also the comprehensive 2010 report by the National Research Council of the National Academy of Sciences that analyzed the effects of biotech crops on farmers and the environment. "Many U.S. farmers who grow genetically engineered (GE) crops are realizing substantial economic and environmental benefits—such as lower production costs, fewer pest problems, reduced use of pesticides, and better yields—compared with conventional crops," notes the study. Although the report does discuss the problem of increasing pest resistance to biotech crops, the development of weeds resistant to herbicides is not a problem peculiar to biotech crops, but is likely exacerbated by the fact that so many biotech varieties incorporate resistance to a single herbicide, glyphosate. The good news is that new varieties are including tolerance to other herbicides. Mixing and matching these crops will better control the development of herbicide-resistant weeds.

Some environmentalist critics claim that genes from genetically modified crops will "contaminate" the natural environment and conventional crops. A 2003 report by the International Council for Science (ICSU) found that "there is no evidence of any deleterious environmental effects having occurred from the trait/species combinations currently available."

Meanwhile, no matter what effects either conventional or GM crops have on biodiversity in crop fields, they pale in comparison to the impact that the introduction of modern herbicides and pesticides sixty years ago had on farmland biology. Thanks to GMOs, farmers' fields became dramatically more productive and comparatively weed- and pest-free.

Biotech Crops and Feeding a Hungry World

In 2009, the Union of Concerned Scientists issued a report, *Failure to Yield,* by chief scientist Doug Gurian-Sherman claiming that modern biotechnology had not increased "intrinsic" crop yields—that is, the highest yield possible under ideal conditions. This assertion is a red herring. Current varieties of biotech crops boost yields chiefly by preventing weeds from using up sunlight and nutrients and insects from destroying them. In other words, biotech crops increase operational yields, the yields actually obtainable in the field taking into account factors such as pests and environmental stresses.

Keep in mind that farmers are not stupid, especially not poor farmers in developing countries. The UCS report acknowledged that American farmers had widely adopted biotech crops in the previous thirteen years. Why? "The fact that the herbicide-tolerant soybeans have been so widely adopted suggests that factors such as lower energy costs and convenience of GE (genetically engineered) soybeans also influence farmer choices," noted the report. Indeed. Surely a UCS advocacy scientist should view saving fossil fuels that emit greenhouse gases as an environmental good. And what does Gurian-Sherman mean by "convenience"? Later, he admits that biotech herbicide-resistant crops save costs and time for farmers. Herbicide resistance is also a key technology for expanding soil-saving no-till agriculture, which, according to a report in 2003, saves 1 billion tons of topsoil from eroding annually. In addition, no-till farming significantly reduces the runoff of fertilizers into streams and rivers.

The UCS report correctly observed, "It is also important to keep in mind where increased food production is most needed—in developing countries, especially in Africa, rather than in the developed world." Which is exactly what is happening with biotech crops in poor countries. Currently, 18 million farmers around the world are planting biotech crops. Notably 90 percent of the world's biotech farmers—that is, 16.5 million—are small and resource-poor farmers in developing countries such as China, India, and South Africa. Gurian-Sherman is right that biotech contributions to yields in developed countries are comparatively modest.

Farmers in the United States and Canada already have access to and can afford to deploy the full armamentarium of modern agricultural technologies, so improvements are going to be at the margins. Nevertheless, it is instructive to compare the rate of increase in corn yields between the biotech-friendly United States and biotech-hostile France and Italy over the past ten years. University of Georgia crop scientist Wayne Parrott notes, "In marked contrast to yield increases in the U.S., yields in France and Italy have leveled off."

Biotech Crops Are Pro-Poor

Yield increases are much greater in poor countries. In 2004, the UN's Food and Agriculture Organization declared that crop biotechnology can be a "pro-poor agricultural technology." The FAO pointed out that crop biotechnology "can be used by small farmers as well as larger ones; it does not require large capital investments or costly external inputs and it is relatively simple to use. Biotechnologies that are embodied in a seed, such as transgenic insect resistance, are scale neutral and may be more affordable and easier to use than other crop technologies."

A 2006 study found that biotech insect-resistant cotton varieties boosted the yields for India's cotton farmers by 44 to 63 percent. Exasperatingly, some anti-biotech activists counter that these are not really yield increases, merely the prevention of crop losses. Of course, another way to look at it is that these are increases in operational yields. Yield increase or crop loss prevention, this success led in 2013 to nearly 90 percent of India's cotton fields being planted with biotech varieties. Similarly, biotech insect-resistant corn varieties increased yields (or prevented losses) by 24 percent in the Philippines.

More recently, a 2010 review article in *Nature Biotechnology* found that "of 168 results comparing yields of GM and conventional crops, 124 show positive results for adopters compared to non-adopters, 32 indicate no difference and 13 are negative." With regard to feeding the world, yield increases are greater for poor farmers in developing countries than for farmers in rich countries. "The average yield increases for developing countries range from 16 percent for insect-resistant corn to 30 percent for insect-resistant cotton," the *Nature Biotechnology* article notes, "with an 85 percent yield increase observed in a single study on herbicide-tolerant corn."

In 2013 the Centre for Environmental Strategy at the University of Surrey published a working paper that looked at the agronomic, environmental, and socioeconomic impacts of biotech crops since they were commercialized in 1996. The researchers found, "Overall, the impact of GM crops has been positive in both the developed and developing worlds." The adoption of biotech crops increased yields and used less energy. "Ecologically, non-target and beneficial organisms have benefitted from reduced pesticide use, surface and ground water contamination is less significant and fewer accidents occur to cause health issues in farm workers," they noted.

The Centre for Environmental Strategy basically confirmed the earlier findings of a 2011 study by four agronomists at the University of Reading. Those researchers reported, "A considerable body of evidence has accrued since the first commercial growing of transgenic crops which suggests that they can contribute in all three traditional pillars of sustainability, i.e. economically, environmentally and socially." With respect to social and economic aspects of sustainable agriculture, the researchers found that the adoption of biotech crops can increase farmers' incomes. "The increase in income to small-scale farmers in developing countries can have a direct impact on poverty alleviation and quality of life, a key component of sustainable development," they noted.

In 2012, two British environmental scientists reviewed the past fifteen years of published literature on the agronomic and environmental effects of biotech crops and found that such crops increase yields and produce impacts that are largely "positive in both developed and developing world contexts." They add, "The often claimed negative impacts of GM crops have yet to materialize on large scales in the field."

Indeed they have not.

Anti-Biotech Activists Kill and Blind Poor Children

Many leading environmental groups are against Golden Rice, a crop that could prevent blindness in half a million to 3 million poor children a year and alleviate vitamin A deficiency in some 250 million people in the developing world. By inserting three genes, two from daffodils and one from a bacterium, scientists at the nonprofit Swiss Federal Institute of Technology created a variety of rice that produces the nutrient beta-carotene, the precursor to vitamin A.

Agronomists at the nonprofit International Rice Research Institute in the Philippines have been crossbreeding the variety, called Golden Rice because of the color produced by the beta-carotene, with well-adapted local varieties and want to distribute the resulting plants to farmers all over the developing world who eat it as a staple food. In 2013, some Filipino "farmers" rampaged through the fields where the IRRI was growing the Golden Rice variety. The "farmers" were later identified as anti-biotech activists who have worked with Greenpeace in the past to block other biotech crop varieties.

Frankly, the scientific community has been far too passive for way too long in confronting the disinformation campaigns of anti-biotech groups such as Greenpeace, Friends of the Earth, and the Union of Concerned Scientists. But the Golden Rice atrocity finally aroused researchers. In August of 2013, *Science* magazine published a strong editorial, "Standing Up for GMOs," condemning activists for their anti-scientific attacks on crop biotechnology.

From the editorial: "If ever there was a clear-cut cause for outrage, it is the concerted campaign by Greenpeace and other nongovernmental organizations, as well as by individuals, against Golden Rice." The scientists pointed out that vitamin A deficiency causes blindness and compromises the body's immune system. The result is blindness for half a million children and 1.9 to 2.8 million preventable deaths annually, mostly of children under five years old and women. The statement notes that environmentalist campaigns are responsible for stalling the release of Golden Rice to farmers for more than a decade.

"Introduced into commercial production over 17 years ago, GM crops have had an exemplary safety record," reads the statement. It adds, "And precisely because they benefit farmers, the environment, and consumers, GM crops have been adopted faster than any other agricultural advance in the history of humanity."

Despite this clear record of safety, the statement continues, "The anti-GMO fever still burns brightly, fanned by electronic gossip and well-organized fear-mongering that profits some individuals and organizations." Hooray! It's great that thousands of researchers have finally called out Greenpeace and other anti-biotech activist groups for peddling and profiting from their disinformation.

In 2014, researchers in Germany and California published another study that calculated that the delay in getting Golden Rice into the fields of poor farmers in India has resulted in the loss of 1.4 million life-years in India that would otherwise have been saved. As the study noted, Golden Rice is "a cost efficient solution that can substantially reduce health costs." Continued environmentalist opposition to this technology is just plain evil.

Biotech Crops and Superpests

Always willing to push its campaign against biotech crops, the Union of Concerned Scientists darkly advises that "genetically engineered crops can potentially cause environmental problems that result directly from the engineered traits." The UCS adds that "the most damaging impact of GE in agriculture so far is the phenomenon of pesticide resistance." The UCS goes on to warn that some insect and plant pests are already becoming resistant to biotech crop protection technologies.

What the UCS and other environmental lobbying groups must know is that researchers have documented for decades the evolution of pests resistant to conventional and organic insecticides and herbicides. It's almost as though the environmentalists had never heard of DDT. This is basic natural selection at work.

Consider, for example, the 1984 article "History, Evolution, and Consequences of Insecticide Resistance," in the journal *Pesticide Biochemistry and Physiology*. Note that the first commercial biotech corn was planted in 1996—that is, twelve years after this article appeared. From the abstract:

> The first inkling of what the future held with respect to
> pesticide resistance of arthropods may be found in 1897
> writings concerning the difficulties of controlling San Jose
> scale (*Quadraspidiotus perniciosus* (Comstock)) and codling
> moth (*Laspeyresia pomonella* (L.)). Eighty-three years later,
> the ever-growing list of resistant species involved 14 orders
> and 83 families, and numbered 428 different insects and
> acarines (e.g., ticks), of which 61 percent are of agricultural
> importance and the remainder of medical/veterinary
> concern.

So what to do to address the emerging pesticide resistance pro[...]
Hit them twice or thrice. Researchers well before the advent of the [...]
tech crop era had devised strategies for slowing down the evolution o[...]
pesticide resistance in insects. For example, a 1989 article entitled "The
Evolution of Insecticide Resistance: Have the Insects Won?" in the journal *Trends in Ecology and Evolution* noted: "A mixture of insecticides . . . can delay the evolution of resistance by several orders of
magnitude compared with a rotation. Mixtures work because insects
that receive a lethal dose of one insecticide are simultaneously dosed
with the other insecticide as well. Only extremely rare individual pests,
which have resistance mechanisms against both chemicals, will survive." Those few that do survive will have difficulty passing along
their double resistance because they will typically breed with untreated
nonresistant mates.

This multiple-treatment regimen is precisely the strategy that plant
breeders use when they "stack" several traits as a way to thwart pest
resistance. Stacking traits is how researchers at Monsanto have created
the Genuity SmartStax crop varieties. The new Monsanto Genuity corn
variety incorporates six different genes aimed at controlling insect pests
plus two for herbicide resistance. In addition, the company is developing Genuity varieties that are drought resistant and use less fertilizer.

What about "superweeds"? Again, the evolution of resistance by
weeds to herbicides is nothing new and is certainly not a problem specifically related to genetically enhanced crops. As of April 2014, the International Survey of Herbicide Resistant Weeds reports that there are
currently 429 uniquely evolved cases of herbicide-resistant weeds globally involving 234 different species. Weeds have evolved resistance to
22 of the 25 known herbicide sites of action and to 154 different herbicides. Herbicide-resistant weeds have been reported in 81 crops in sixty-
five countries.

A preliminary analysis by University of Wyoming weed scientist
Andrew Kniss parses the data on herbicide resistance from 1986 to 2012.
He finds no increase in the rate at which weeds become resistant to her-
bicides after biotech crops were introduced in 1996. Since Roundup

most popular herbicide used with biotech crops, have

l species resistant to Roundup increased? Kniss finds

it of Roundup-resistant weeds has occurred more

on-biotech crops. "Glyphosate-resistant weeds

ate use, not directly due to GM crops," he points

cide resistant weed development is not a GMO problem, it

a herbicide problem."

Biotech Crops and "Contamination"

Another environmentalist attack on biotech crops is the assertion that GMO pollen "contaminates" organic crops and thus economically harms organic farmers. Not so, says University of Oklahoma law professor Drew Kershen. First, Kershen points out that organic standards are process standards, not product standards. Organic crops receive certification because of the way they are produced: no chemical fertilizers, no synthetic pesticides, and so forth. Saying a product is organic does not mean it is totally free of chemical fertilizers and synthetic pesticides. Organic farmers already experience accidental pesticide drift and the admixture of conventional seeds. They can still obtain organic certification, provided they conscientiously follow all the rules specified by the federal government's National Organic Program. The same standard could easily apply to organic farmers whose crops have experienced minor interbreeding with transgenic crops.

Second, Kershen notes that US law generally does not allow those with special sensitivity to an activity to declare that they have been harmed by it. It is their responsibility to protect themselves from the activities they dislike. "You do not have a claim based on your assertion of increased sensitivity," Kershen explains. "If you don't like to hear rock music, you can't prohibit your neighbors from playing it at reasonable levels. You have to protect yourself. Stay away from concerts. Soundproof your home." Similarly, organic farmers could perhaps grow borders that would insulate their crops from their neighbors' pollen flow.

But won't consumers of organic products reject a farmer's crops if they think it is "contaminated" with genes from genetically enhanced crops? Kershen says that argument won't wash either. He offers an example in which a tattoo parlor legally opens between a florist and a Christian bookstore and advertises a special on satanic tattoos. Customers offended by the tattoo shop begin avoiding the florist and bookstore. Under American common law, the florist and the bookstore do not have a cause of action, because "economic expectation is not recoverable." Similarly, an organic farmer who expected to sell his crop at a premium would nevertheless be able to sell it at market rates as a conventional crop; he loses only the premium he expected to gain.

Kershen makes the further point that those who have created a niche market should be the ones responsible for protecting it. "After all, they are the ones trying to differentiate their products in order to obtain higher profits," he notes. "Therefore, the rest of us who don't care shouldn't be saddled with the costs of defending their self-imposed standards and labeling." That would be akin to forcing conventional meat packers to carry "nonkosher" labels on all their meats for the benefit of kosher meat producers.

One method of dealing with the question of pollen flow is for organic farmers to set reasonable tolerances. Many activists and organic farmers advocate "zero tolerance" standards that would in effect outlaw genetically enhanced crops. Since every independent scientific body that has ever looked into the safety of biotech crops has found them to be safe, this would be an absurd requirement.

Crops have exchanged pollen for millennia, and they will continue to do so. Seed breeders have decades of experience in setting tolerances for seed purity. As Mark Condon, vice president for international marketing at the American Seed Trade Association, has explained, "seed purity certification standards were commonly set at 98 percent to 99 percent varietal purity levels or a standard of 1 percent to 2 percent adventitious genetic impurity." It should be possible to maintain similar standards for organic crops. Since organic farmers set their own standards, they could easily adopt these tolerances and save themselves and conventional farmers a lot of trouble.

Another possibility is that biotechnology itself may come to the aid of organic farmers. Already researchers have developed a way to make genetically enhanced crops sterile, which would limit the spread of transgenes. It might turn out that such technical fixes will make biotech farmers the low cost avoiders and thus shift the balance in favor of organic farmers' claims. You'd think organic farmers and environmentalists would be clamoring for these gene-sequestering technologies, but instead they strenuously oppose them. Why?

One amusing and particularly telling argument for opposition to this technology comes from Vandana Shiva. "The possibility that [biotech sterility] may spread to surrounding food crops or to the natural environment is a serious one," Shiva gloomily posits in her book *Stolen Harvest*. "The gradual spread of sterility in seeding plants would result in a global catastrophe that could eventually wipe out higher life forms, including humans, from the planet." Really? This dire scenario is not just implausible but biologically impossible: *The gene technology causes sterility; that means, by definition, that the sterility can't spread.*

Biotech Seed Companies Do *Not* Sue Farmers for Accidental Pollination

Why are there patents in the first place? As Abraham Lincoln (himself a patent holder) explained in 1858, patents add "the fuel of *interest* to the *fire* of genius, in the discovery and production of new and useful things." Patents are also a disclosure mechanism in which inventors are awarded exclusive use of their inventions for twenty years in exchange for clearly revealing to the rest of us how they are made, thus avoiding a world encumbered by trade secrets.

Activists have vigorously pushed the notion that biotech seed companies regularly and aggressively sue farmers who accidentally acquire patented genetically engineered germplasm via pollen drifting onto their farms from neighboring biotech crops. Their chief villain in this anti-biotech morality tale is nearly always the biotech seed company Monsanto. Such a situation would obviously be unfair and wrong. Fortunately, it has never happened.

Take the notorious case of Saskatchewan canola farmer Percy Schmeiser. Activists made a hero out of Schmeiser, who was sued by Monsanto for sowing its variety of herbicide-resistant canola without paying for it. Schmeiser claimed that the canola seed had somehow blown across his fence line. That is not a problem for Monsanto; accidents happen. How did Schmeiser find out that this had supposedly happened? He sprayed three acres of canola underneath some power lines with some Roundup and the plants didn't die. Later testing showed that they were in fact a Monsanto herbicide-resistant variety.

In its decision, the Supreme Court of Canada outlined what then happened. "Mr. Schmeiser complained that the original plants came onto his land without his intervention. However, he did not at all explain why he sprayed Roundup to isolate the Roundup Ready plants he found on his land; why he then harvested the plants and segregated the seeds, saved them, and kept them for seed; why he next planted them; and why, through this husbandry, he ended up with 1,030 acres of Roundup Ready Canola which would otherwise have cost him $15,000," wrote the court. Given these circumstances, it is not surprising that the court decided against Schmeiser.

In a more recent case, Indiana soybean farmer Vernon Hugh Bowman admitted that he had saved Monsanto soybeans and replanted them. He claimed that once he bought the seeds, what he did with them afterward, including saving and replanting, did not violate Monsanto's patents. Bowman had the option of planting conventional soybeans whose seeds he could legally save for replanting each year. It's clear that he chose not to do this because he specifically wanted the weed-control convenience Roundup Ready seeds afforded him; he just didn't want to pay for them.

The case of *Bowman v. Monsanto Co.* made it to the US Supreme Court in 2013. During oral argument, Chief Justice John Roberts got right to the heart of the matter by asking: "Why in the world would anybody spend any money to try to improve the seed if as soon as they sold the first one anybody could grow more and have as many of those seeds as they want?" The court ruled unanimously against Bowman.

Given the hullabaloo spread by activists, one might be forgiven for thinking that big biotech companies are suing farmers all of the time. That's not so. In fact, the vast majority of farmers in the United States keep their promise not to save seeds from crop varieties they purchase from biotech companies. Consider Monsanto's record. The company sells its seeds to about 250,000 American farmers every year. Between 1997 and 2012, Monsanto filed only 145 lawsuits over seed patents, of which only 9 actually went to trial. Monsanto won each case at trial. An additional 700 or so cases have been settled out of court.

GMO Labeling?

Don't consumers have the right to know that they are eating foods made with ingredients from biotech crops? Concerned consumers should already know that in the United States, somewhere around 80 percent of all prepared foods in grocery stores contain ingredients from biotech crops. What sort of ingredients? Mostly soybean oil and corn syrup, in which biotech proteins and DNA are very nearly undetectable.

If consumers don't know this, perhaps it means that they are not all that concerned. Nevertheless, polls regularly find that large majorities favor labeling biotech foods. There are good reasons to suspect that this may be because, after all, what self-respecting person ever would tell a pollster that he or she *didn't want to know* something about an apparently important issue? A 2010 European Commission report, however, maintains that polls may not be a good way to evaluate actual consumer attitudes toward foods made with biotech crops. The researchers report that despite strongly negative polls, when it came to looking at actual buying behavior, "most people do not actively avoid

GM food, suggesting that they are not greatly concerned with the GM issue."

Under the specious banner of consumer rights, an unholy alliance of donation-hungry fear-mongering environmental groups and profit-hungry organic foods companies in the United States is campaigning for mandatory labeling of food made using ingredients from biotech crops. In several states, ballot initiatives have come up for a vote, and in others, legislatures are considering such a mandate. So why not label the foods and be done with it? Because the opponents of biotech farming are not actually interested in trying to give consumers information. Instead, their fond hope is that consumers will mistake such labels as warning labels and then shy away from buying foods made from biotech crops. In other words, they aim to spread disinformation, not provide real information.

This is why the US Food and Drug Administration has refused to require labels on foods made from biotech crops. As has been shown earlier, every independent scientific body that has considered the issue has determined that the current varieties of biotech crops are safe to eat. The FDA states that the agency "requires special labeling . . . in cases where the absence of such information may pose special health or environmental risks." The FDA further notes that it "has no basis for concluding that bioengineered foods differ from other foods in any meaningful or uniform way, or that, as a class, foods developed by the new techniques present any different or greater safety concern than foods developed by traditional plant breeding." Since foods using biotech ingredients do not pose special health or environmental risks, they need not be labeled. A statement issued in 2012 by the board of directors for the American Association for the Advancement of Science noted, "The FDA does not require labeling of a food based on the specific genetic modification procedure used in the development of its input crops. Legally mandating such a label can only serve to mislead and falsely alarm consumers." Misleading and alarming is the goal of activists.

In a March 14, 2013, editorial, "Why Label Genetically Engineered Foods?" *The New York Times* pointed out that there is no scientific reason for requiring such labels. The *Times* editorial made the sensible observation that if people wanted to avoid foods made with biotech ingredients, they could simply buy foods labeled organic.

In 2013, the grocery chain Whole Foods Market announced that it will require its suppliers to either source non-GMO ingredients or to clearly label products with ingredients containing GMOs by 2018. Of course, private companies have a perfect right to make such contractual arrangements.

In April 2014, Representative Mike Pompeo (R-KS) and Representative G. K. Butterfield (D-NC) introduced the Safe and Accurate Food Labeling Act. The act would require the FDA to promulgate rules with regard to voluntarily labeling foods as containing or not containing ingredients from bioengineered crops. The new FDA labels "may not suggest either expressly or by implication that foods developed without the use of bioengineering are safer than foods produced from, containing, or consisting of a bioengineered organism," and vice versa. In addition, the act would preempt individual state biotech labeling mandates. Naturally, this preemption provision has riled anti-biotech activists. Given the vexed politics of the labeling fight, it is doubtful that this act will become law.

University of California agronomist David Zilberman worries that compulsory labeling would stigmatize biotech products and reduce investment in new GM traits. "The net effect will be to slow the development of agricultural biotechnology, and this in turn may negatively affect health, the economy, and the environment," he writes. "It is actually counterproductive to the many environmental and social goals that we cherish. Therefore, labeling of GMOs will be a step in the wrong direction."

Looking at the political trajectory of the battle over labels, it is probable that food companies will end up simply slapping a label on everything saying: "This product may contain ingredients from modern biotech crops." The information should merely be included in the list of ingredients. One can hope that as the EU report suggests, for most consumers, including those shopping at Whole Foods, such information will pretty soon go in one eye and out the other with little effect on their purchasing decisions.

Biotech Corporate Domination

In developed countries, the production of biotech crop varieties is currently dominated by six seed companies. Anti-biotech activists make much of this corporate concentration, slyly suggesting that food security is somehow threatened by corporate control. It didn't have to be this way. This corporate concentration is the predictable result of the restrictive regulatory system promoted by the activists themselves. Prior to the 1990s, there were scores of seed companies in the United States alone. As modern biotech crop seeds were being developed, environmentalists insisted that each new variety had to undergo separate approvals, even though seed companies were enhancing crops using essentially the same set of traits—that is to say, pest and herbicide resistance. This is wholly unnecessary since, if due care is observed, a trait that is safe in one crop or variety is safe in another.

Environmentalists hoped that this onerous procedure of regulatory approval for each individual variety would make biotech crops uneconomic. Instead, it drove most of the small seed companies out of business. Today it costs about $136 million, in part because of regulatory requirements, to bring a new bioenhanced variety to market. Small companies and nonprofit researchers cannot afford this process. BASF Corporation, Bayer CropScience, Dow AgroSciences, DuPont/Pioneer Hi-Bred, Monsanto, and Syngenta have the financial and legal resources to navigate the regulatory maze. These companies have a great deal to thank environmentalists for, since the regulations the activists demanded have significantly enhanced their corporate bottom lines by blocking competition.

The Farmer Suicide Myth

There is apparently no evil of which crop biotechnology cannot be accused by anti-biotech campaigners, up to and including "genocide." One of the more egregious examples of outright falsehoods being peddled by activists is that the introduction of modern pest-resistant cotton in India provoked hundreds of thousands of farmer suicides. In 2002, a local Indian seed company made available bioenhanced cotton seeds using Monsanto's Bt insect resistance trait. Farmers flocked to it.

Activist-charlatan Vandana Shiva has been a particularly enthusiastic hawker of the "suicide seeds" accusation. In 2006, her institute issued a report that asserted that Bt cotton had failed, driving poor farmers into debt and despair, from which tens of thousands sought escape by killing themselves. "Genetic engineering is killing Indian farmers," starkly concludes the report. As recently as 2013, Shiva has insisted, "Two hundred and seventy thousand Indian farmers have committed suicide since Monsanto entered the Indian seed market. That's more than a quarter-million. It's a genocide." Shiva's charges have been the subject of numerous credulous articles and documentaries. They are completely false.

In a comprehensive analysis, "The GMO-Suicide Myth," published in 2014 in *Issues in Science and Technology*, science journalist Keith Kloor eviscerates Shiva's stories, revealing her as the liar that she is. First, since 2002, Indian farmers have eagerly embraced Bt cotton. Ninety percent of India's cotton farmers now grow Bt cotton. As a consequence, farmers use much less pesticide and yields have soared. Kloor reports that India's agricultural minister said in 2012 that the country "has harvested an average of 5.1 million tons of cotton per year, which is well above the highest production of 3 million tons before the introduction of Bt cotton."

And more than that, Bt cotton has dramatically improved farmers' incomes and consequently boosted their food security. In 2013, two German university researchers reported that "the adoption of GM cotton has significantly improved calorie consumption and dietary quality, resulting from increased family incomes. This technology has reduced food insecurity by 15–20% among cotton-producing households." Farmers are not stupid. They plant only crops that they think will benefit them.

It is true that far too many Indian farmers have committed suicide. So what is really going on? Recent research by University of Syracuse political scientist Anoop Sadanandan finds that many Indian farmers commit suicide largely as the result of changes to banking made by the Indian government that have had the side effect of throwing them onto the mercies of local loan sharks. "There is, however, no evidence to suggest that the cultivation of a particular crop was related to suicides in India," reports Sadanandan. "For instance, I did not find a systematic relation between cotton cultivation, which is often linked to farmer suicides in the country, and farm suicide rates." A recent preliminary analysis by University of Manchester social statistician Ian Plewis concludes that "the available evidence does not support the view that farmer suicides have increased following the introduction of Bt cotton." Plewis adds, "Taking all [Indian] states together, there is evidence to support the hypothesis that the reverse is true: male farmer suicide rates have actually declined after 2005 having been increasing before then." He also notes that the farmer suicide rate in India is similar to the rates among farmers in Scotland and France.

"Blaming farmer suicides on Bt cotton thus seems to be not only incorrect but also a distraction from the real causes of a tragic problem," concludes Kloor. "One is left wondering what problem Vandana Shiva and other like-minded activists are actually interested in solving, since it does not seem to be the livelihoods of Indian farmers." Indeed.

Future Biotech Developments

The benefits of biotechnological progress are so overwhelming that it is unlikely that environmentalist opposition can entirely stop it. So let's take a brief look at what biotech researchers are currently trying to bring to consumers.

Researchers are using biotechnology to endow trees with useful qualities. For example, ArborGen has created a fast-growing freeze-tolerant eucalyptus tree that can flourish on tree plantations in the southern United States. Eucalyptus grows very fast, and its wood is used to produce pulp and paper. It produces about seventeen tons of wood per acre per year and is harvested after seven years. Natural hardwood forests produce about two tons of wood per year and are harvested after forty years of growth. In addition, the biotech eucalyptus is engineered to produce no pollen or seeds so that it cannot escape into the wild.

Consider, too, that choosing to plant a conventional poplar or, say, a poplar genetically modified to produce less lignin will have far fewer ecological effects than choosing between planting a poplar, modified or not, and a conifer species. "The specific changes in wood chemistry imparted by GM [genetic modification] will be orders of magnitude less than the vast number of new chemicals that distinguish a pine from an aspen," notes a 2001 Oregon State University study. In other words, planting pine trees would have much greater effects on soil, pests, water retention, and biodiversity than the comparatively minor effects entailed by choosing to plant either conventional poplars or modified ones. Not bad effects, just different ones.

Anti-biotech activists unaccountably miss the crucial point that improving the productivity of trees grown on plantations reduces the pressure to cut down trees in wild forests. Resources for the Future analyst Roger Sedjo estimates that most of the world's wood products could be derived from tree plantations occupying about 7 percent of the world's currently forested area. Improving the productivity of tree plantations by means of biotechnology would shrink that area even further. Cultivating trees on plantations spares land for natural forests to grow, and the more productive the trees, the more the land that can be spared for nature.

Biotechnology could also restore the majestic American chestnut to its home in the forests of the eastern United States. In the early twentieth century, chestnut blight fungus introduced from overseas devastated American chestnuts—the then-dominant trees in eastern forests stretching from Maine to Mississippi. Researchers at the State University of New York College of Environmental Science and Forestry (SUNY), working with the American Chestnut Research and Restoration Project, used genetic engineering techniques to introduce a gene from wheat that confers strong blight resistance into American chestnut trees. They recently ceremonially planted some of the new blight-resistant chestnuts at the New York Botanical Garden in the Bronx, where the blight was first identified. Nevertheless, some activists have denounced these trees as "wheat-gene tweaked freaks."

Currently crops typically use only 30 to 50 percent of the nitrogen fertilizer they receive. Nitrogen fertilizer runoff contributes to water pollution and is the primary source of anthropogenic nitrous oxide, a greenhouse gas that is 300 times more potent than carbon dioxide. Agriculture contributes up 12 percent of man-made global warming emissions. So one would think that environmentalists would welcome Arcadia Biosciences's new biotech variety of rice that needs 50 to 60 percent less nitrogen fertilizer than conventional varieties, but they haven't. The really good news is that research into transferring this same set of fertilizer-thrifty genes into other crops is moving rapidly forward. For example, Arcadia has created a biotech canola that uses two-thirds less nitrogen fertilizer but yields the same as conventional varieties. This application of biotechnology to increase the efficiency of nitrogen use not only protects the environment but also reduces the amount that farmers have to pay for fertilizer.

Even more remarkably, Arcadia Biosciences announced in 2014 successful field trials of a biotech rice variety that combines salt tolerance, drought tolerance, and nitrogen use efficiency. The salt tolerance is derived from thale cress, the drought tolerance from a common soil bacterium, and the nitrogen use efficiency from barley. To state the obvious, there is no way that conventional breeding could have produced this type of rice. This "super-rice" will be particularly advantageous to farmers in developing countries, who must cope with poor soils, spotty rains and no irrigation, and high fertilizer prices.

Another promising area of research being done at the nonprofit International Rice Research Institute in the Philippines involves using genetic engineering to transfer the C4 photosynthetic pathway into rice, which uses the less efficient C3 pathway now. This could boost rice yields tremendously, perhaps as much as 50 percent, while reducing water use. In addition to salt and drought tolerance, researchers are pursuing all manner of other ways to boost crop production, including heat tolerance, along with viral, fungal, and bacterial disease resistance.

For example, Danforth Center is working on a biotech cassava that can resist the viral cassava mosaic disease that destroys a third of Africa's cassava crop annually. One of the first great successes of crop biotechnology was engineering resistance to ringspot virus into Hawaiian papayas back in the 1990s. Recently anti-biotech activists in Hawaii have taken to hacking down papayas in the middle of the night. Researchers created a biotech wheat that reduced losses from powdery mildew fungus by as much as 30 percent. The nonprofit Swiss research was halted when anti-biotech activists vandalized the test fields. British researchers endowed a variety of potato to resist infection by a relative of the same pest that caused the Irish potato famine. The new bioenhanced potato cut losses by between 50 and 75 percent. If these biotech techniques pan out, they could improve crop productivity and thus reduce agriculture's call on land, water, and air resources.

Exciting biotech research is not confined just to plants. Instead of turning a third of the American corn crop into biofuels, biotech researchers are working to generate algae that can suck carbon dioxide out of the air and secrete fuels. Canadian researchers at the University of Guelph have created the "enviropig." Combining genes from a common bacterium with those from mice, the researchers have endowed the enviropigs with the ability to digest phosphorus from plants. Ordinary pigs cannot digest about 75 percent of the plant phosphorus in their feed. The new biotech trait boosts their growth and reduces the amount of phosphorus in their manure.

AquaBounty Technologies, which is based in Waltham, Massachusetts, has developed a fast-growing biotech salmon. AquaBounty's salmon eat 10 to 25 percent less feed than do conventional Atlantic salmon. AquaBounty salmon grow twice as fast as wild Atlantic salmon as the result of the installation of two genes, a promoter gene derived from the eel-like ocean pout and a growth hormone gene from Chinook salmon. People have long eaten both species. The fish would be raised onshore in physically contained facilities so that there is no chance they could escape.

The company has been seeking FDA approval since 1995. Finally, in 2012, the agency's Veterinary Medicine Advisory Committee found no biologically relevant differences between conventional and biotech salmon on nutrition and allergenicity. Consequently, the committee concluded, "With respect to food safety, FDA has concluded that food from AquAdvantage Salmon is as safe as food from conventional Atlantic salmon." Naturally, environmental organizations oppose this technology even though most of the world's wild fisheries are woefully overexploited. And adding insult to injury, members of Congress from states with fishers of wild salmon who fear competition introduced legislation to ban the biotech salmon.

Biotechnologists are also developing resource-thrifty ways to produce foods and ingredients through fermentation using genetically modified yeasts and bacteria. For example, the California start-up Muufri (pronounced "moo free") is engineering yeast to produce the components of cow's milk. The company hopes to bring its animal-free milk to market by 2017. In conventional dairy production, it takes about a thousand liters of water to produce one liter of milk, and the waste generated by one cow equals that of twenty to forty people. In addition, vast amounts of land are devoted to growing pasture, alfalfa, and corn as feed for cows. On a global basis, dairy cattle emit 17 percent of the greenhouse gases produced by livestock. If Muufri milk finds favor with consumers, the environmental benefits are potentially enormous.

The Biotech Farmer and the Organic Environmentalists Should Be Friends

"The farmer and the cowman should be friends," sing the folks in the 1943 Rodgers and Hammerstein musical *Oklahoma!* So why shouldn't biotech and organic growers be friends, too? After all, both want to make decent livings as they produce nutritious food for a hungry world while enhancing soil fertility; fending off diseases, pests, and weeds; reducing costly and environmentally damaging inputs; preventing erosion and fertilizer runoff; conserving water; and protecting biodiversity. Some environmentalists are arguing that the world should "freeze" the amount of cropland at current levels and produce food for the 9 billion people living in 2050 only from the acres already being cultivated. Since organic agriculture as currently practiced produces on average about 75 percent of the yields of conventional farming, a global switch to organic farming would make it impossible to freeze cropland for today's population, much less for 9 billion in four decades.

On the other hand, high-yielding biotech seeds combined with organic soil management techniques could deliver just the sort of "sustainable intensification" that humanity needs. Organic soil management recycles nutrients, increases organic content, and enhances moisture retention. Adding crop varieties that can generate their own fertilizer, resist drought, flourish in saline soils, fight off diseases and pests without chemical sprays, and grow without weeding and plowing makes organic agriculture that much gentler on the natural environment.

In their 2008 book *Tomorrow's Table: Organic Farming, Genetics, and the Future of Food*, the wife-and-husband team of University of California at Davis plant geneticist Pamela Ronald and organic farmer Raoul Adamchak write, "We believe that the judicious incorporation of two important strands of agriculture—genetic engineering and organic farming—is key to helping feed the growing population in an ecologically balanced manner." Sounds about right.

6

Can We Cope
with the Heat?

IN 2005, I PUBLICLY CHANGED MY MIND ABOUT climate change. I concluded that the balance of the scientific evidence indicated that man-made global warming likely posed a significant problem for humanity. My new assessment did not please a number of my public policy friends, some of whom made their disappointment clear. Perhaps the most amusing but nevertheless painful episode occurred during the 2007 annual gala dinner of the Competitive Enterprise Institute, a Washington, D.C.–based free-market think tank.

The master of ceremonies was former *National Review* editor John O'Sullivan. To entertain the crowd, O'Sullivan put together a counterfeit tale in which I ostensibly had given a recent lecture on environmental trends pointing out that most were positive. After the lecture, O'Sullivan told the audience, a young woman supposedly approached me to express her displeasure with regard to my change of mind on climate change. Continuing his fable, O'Sullivan recounted to the hundreds of diners that I had tried to explain why my views had shifted. Eventually realizing that the young woman was having none of it, I then purportedly asked her if it wasn't enough that we two actually agreed on most environmental policy issues. The young woman paused for a moment, said O'Sullivan, and then retorted, "I suppose that Pontius Pilate made some good decisions, too." Being compared, even in jest, to the Roman governor who consented to the crucifixion of Jesus is, to say the least, somewhat disconcerting.

Welcome to the most politicized science of our time. As such, sorting through the claims and counterclaims of assorted "alarmists" and "deniers" with regard to the scientific research and policy prescriptions is a fraught undertaking.

On the catastrophe side stands former U.S. vice president Al Gore, who has warned Congress that man-made global warming is "a true planetary emergency." Joining Gore is environmentalist Bill McKibben, founder of the 350.org activist group, who promotes the goal of reducing atmospheric concentrations of carbon dioxide to back below 350 parts per million (ppm) from the current 400 ppm. In his 2014 book, *Oil and Honey,* McKibben sees future climate change as portending "an endless chain of disasters that will turn civilization into a never-ending emergency response drill." McKibben's prescription is a turn away from global consumerism toward the organic and local, to "a nation of careful, small-scale farmers who can adapt to the crazed new world with care and grace, and who don't do much more damage in the process." Fierce progressive activist Naomi Klein in her newest screed, *This Changes Everything: Capitalism vs. the Climate,* declares, "Our economic system and our planetary system are now at war." Klein asserts that the progressive values and policies she advocates are "currently being vindicated, rather than refuted, by the laws of nature." Climate science, she further claims, has given progressives "the most powerful argument against unfettered capitalism" ever. McKibben called Klein's book "the key new text on climate change."

Across the ideological divide, the criers of climate crisis are passion-ately opposed by a cohort of skeptics who reject the claims of impending climate catastrophe, arguing that they are largely based on corrupted and politicized science. One leading voice challenging climate doom is Senator James Inhofe (R-OK), who is now the chairman of the US Senate Committee on Environment and Public Works. Inhofe declared in 2003 that "man-made global warming is the greatest hoax ever perpetrated on the American people." Another of the fiercest challengers of orthodox climate science is Marc Morano, the publisher of the influential Climate Depot website. "The scientific reality is that on virtually every claim—from A-Z—the claims of the promoters of man-made climate fears are failing, and in many instances the claims are moving in the opposite direction," asserted Morano in congressional testimony. He added, "The global warming movement is suffering the scientific death of a thousand cuts." Fred Singer, former director of the US National Weather Service's Satellite Service Center and now head of the Science and Environmental Policy Project, asserts, "There would be very little public interest in funding climate science, were it not for an assertion by alarmists in the political and environmental communities that a human-caused global warming crisis exists." Despite the billions spent on climate research, Singer argues, "There is no convincing evidence that it has been warming."

I have been reporting on climate change for a quarter of a century. I covered the 1992 Earth Summit in Rio de Janeiro, Brazil, at which the United Nations Framework Convention on Climate Change was negotiated. I have since reported from ten of the United Nations annual Climate Change conferences. The anecdote at the beginning of this chapter reveals that after years of reporting on the subject, attending scientific conferences, talking with scientists, and extensively reading the research literature, I have concluded that the balance of the evidence indicates that climate change could become a significant problem for humanity as the twenty-first century unfolds.

First, let's look at what the best scientific evidence is (and is not) telling us about the likely trajectory of future climate change. This includes future temperature increases, sea level rise, shifts in the amounts of snow and rain, and ocean acidification. Next, we will analyze the possibilities of adapting to a changing climate and ask what we owe to future generations. Third, we will parse the costs and benefits of policies proposed to prevent or slow future warming. Fourth, we will probe the sorry history of global climate negotiations. Finally, we examine how human ingenuity can most likely deliver the energy technologies that will address and solve the climate problem well before the end of this century.

How Hot Is It?

According to the reckoning of various international meteorological organizations, 2014 was either the hottest year or close to the hottest year since fairly accurate instrumental temperature records started being kept in the mid-nineteenth century. For example, in January 2015 the US National Oceanic and Atmospheric Administration (NOAA) reported that there is 38 percent chance that 2014 was warmer than 2010 or 2005, the next two warmest years in the NOAA records. The independent climate research group Berkeley Earth also concluded that 2014 was nominally the warmest since the global instrumental record began in 1850 while noting, however, that within the margin of error, it is tied with 2005 and 2010. The UK Met Office and the Climatic Research Unit at University of East Anglia ranked 2014 as tied with 2010 for the warmest year in the record, but added that the uncertainty ranges mean it's not possible to definitively say which of several recent years was the warmest.

Climatologists at the University of Alabama in Huntsville have been tracking global temperatures for the past thirty-six years using satellite data that measure the bottom five miles of the atmosphere. They reported that 2014 was the third warmest year in that record. As analyzed by Remote Sensing Systems, 2014 was only the sixth warmest year in the satellite record. For both satellite data sets, 1998 is the hottest year. All data sets agree that the last ten years or so have been the warmest period during the instrumental record.

What the Science Says

The amount of greenhouse gases in the atmosphere is increasing; the world has warmed; glaciers are melting; and the seas are rising. These facts are not scientifically in dispute. As *Climate Change 2013: The Physical Science Basis,* the 2013 report from the United Nations Intergovernmental Panel on Climate Change (IPCC), states: "Warming of the climate system is unequivocal, and since the 1950s, many of the observed changes are unprecedented over decades to millennia." The report adds, "Each of the last three decades has been successively warmer at the Earth's surface than any preceding decade since 1850." These findings were restated and bolstered in November 2014 in the IPCC's *Climate Change 2014: Synthesis Report.* The vast majority of climate researchers agree that man-made global warming is now under way.

The amounts of carbon dioxide, methane, and nitrous oxide, gases that tend to warm the atmosphere (greenhouse gases or GHG), are at levels unprecedented in at least the last 800,000 years. Atmospheric carbon dioxide is increasing chiefly because humanity releases more as it burns fossil fuels and cuts down forests. Methane is released through natural gas drilling and the flatulence of growing livestock herds. And nitrogen oxide rises as a side effect of fertilizer use and emissions from vehicles. Carbon dioxide concentrations are 40 percent higher than during preindustrial times back in 1750. Thirty percent of the carbon dioxide emitted by human activities has dissolved into the oceans, where it has increased the acidity of the water by 26 percent.

Since the 1880s, the planet has warmed by an average of 0.85°C (1.5°F). The IPCC report notes that since 1951 average global temperature has been increasing at a rate of 0.12°C (0.22°F) per decade. "It is *extremely likely* that human influence has been the dominant cause of the observed warming since the mid-20th century," states that report. In November 2014, the IPCC's *Synthesis 2014 Report* stated, "It is *extremely likely* that more than half of the observed increase in global average surface temperature from 1951 to 2010 was caused by the anthropogenic increase in greenhouse gas concentrations and other anthropogenic forcings together."

In addition, most mountain glaciers and the ice sheets that cover Antarctica and Greenland are melting, and the extent of spring snow cover in the Northern Hemisphere has been declining since 1967. The area of Arctic Ocean summer sea ice has been falling at a rate of between 9.4 and 13.6 percent per decade since 1979. Between 1901 and 2010, sea level rose at a rate of 1.7 millimeters (0.7 inch) per year, increasing average sea level by 0.19 meters (about 8 inches) over that period. As more ice on land drains into the warming oceans, the IPCC estimates that sea level is now rising at 3.2 millimeters (0.12 inches) per year. At this rate, global average sea level would rise by about 0.27 meters (11 inches) by 2100.

In trying to discern how future climate change will play out over the remainder of this century, the IPCC scientists rely on the outputs from computer climate models. As part of the process of computer modeling, they have set up scenarios called Representative Concentration Pathways (RCPs). Each RCP more or less corresponds to certain specified levels of carbon dioxide concentrations in the atmosphere that might be reached by 2100. Currently, the level of greenhouse gases in the atmosphere tallies at about the equivalent of 400 parts per million (ppm) of carbon dioxide, up from 280 ppm in 1750. For example, RCP2.6 reaches 475 ppm carbon dioxide equivalent; RCP4.5 is up to 630 ppm; RCP6 grows to 800 ppm; and RCP8.5 rises to 1313 ppm.

The IPCC reports, "Global surface temperature change for the end of the 21st century is likely to exceed 1.5°C [2.7°F] relative to 1850 to 1900 for all RCP scenarios except RCP2.6. It is likely to exceed 2°C [3.6°F] for RCP6 and RCP8.5, and more likely than not to exceed 2°C [3.6°F] for RCP4.5." The 2°C number is significant since in ongoing negotiations under the United Nations Framework Convention on Climate Change the agreed international goal is to keep the future average global temperature increase below that threshold. It bears noting that the IPCC reports that "the total increase between the average of the 1850–1900 period and the 2003–2012 period is 0.78°C [1.4°F]." More generally, the computer models on which the IPCC relies project by 2100 that the average increase in global temperature above the 1850–1900 average will be 1°C, 1.8°C, 2.2°C, and 3.7°C, for the RCP2.6, RCP4.5, RCP6, and RCP8.5 scenarios, respectively.

The IPCC reports that computer models forecast that average sea level rise by 2100 will range from 0.26 to 0.98 meters (10 to 38 inches). Two scientific studies published in May 2014 indicated that the glaciers of the West Antarctic Ice Sheet are beginning to melt and so will likely push the estimates of sea level rise by the end of this century toward the higher end of that spread. In an effort to set a highly unlikely worst-case upper limit on possible sea level rise, a study in the October 2014 issue of *Environmental Research Letters* reckoned that sea level could at worst rise by no more than about 6 feet during this century. In another October 2014 analysis, a team of researchers led by Australian National University earth scientist Kurt Lambeck indicated that global average sea level has begun steadily rising in the past 150 years, after having fluctuated no more than 6 to 8 inches during the past 6,000 years. At the Earth's northern pole, the IPCC expects significant reductions in Arctic sea ice extent by 2100 ranging from a low of 43 percent to a high of 94 percent during the annual summer melt. Spring snow cover is projected to decrease in the Northern Hemisphere by 7 to 25 percent by 2100.

The IPCC *Physical Science* report also sets out essentially a "carbon budget" that delineates what quantity of additional greenhouse gases can be emitted without exceeding the 2°C (3.6°F) limit. The IPCC estimates that cumulative carbon dioxide emissions amounting to 5760 gigatonnes (one gigatonne equals one billion metric tons) would mean that there is less than a 33 percent chance of remaining below 2°C. Emitting 4440 gigatonnes raises that to a fifty-fifty chance and emitting even less at 3670 gigatonnes boosts the odds of staying below 2°C rises to 66 percent. Since preindustrial times, humanity has already emitted about 1890 gigatonnes of carbon dioxide. In other words, if the goal is to remain below the 2°C threshold, the IPCC reckons that humanity has already used up about half of its carbon dioxide budget.

In November 2014, the IPCC's *Synthesis Report* concluded, "Scenarios that are *likely* to maintain warming at below 2°C are characterized by a 40% to 70% reduction in GHG emissions by 2050, relative to 2010 levels, and an emissions level near zero or below in 2100."

Back in his 2007 testimony before a congressional committee, former US vice president and prominent climate change crusader Al Gore sought to summarize the consensus view of climate scientists by declaring, "The science is settled." US EPA administrator Lisa Jackson reiterated that view in her 2010 testimony to Congress: "The science behind climate change is settled, and human activity is responsible for global warming." As outlined above, there is indeed broad agreement among climate researchers with regard to a lot of the climate change data and forecasts. Nonetheless, many researchers acknowledge there are some significant unresolved scientific questions that remain to be settled. Further research on these outstanding issues might yield happy results that tend to lower the more dire estimates of future climate change. Let's turn to those questions now.

The Science Is Settled—The Global Temperature Hiatus

"The 5-year running mean of global temperature has been flat for the past decade," admitted a 2013 global temperature update from the NASA Goddard Institute for Space Studies (GISS) at Columbia University, one of the world's leading climate change research centers. At the time of the GISS update, the hiatus in the increase of global average temperature had actually been going on for nearly sixteen years, ever since 1998. In an August 2014 article, longtime IPCC critic Ross McKitrick applied advanced statistical techniques to global temperature data sets that extended the period during which the global temperature has been essentially flat to encompass the past nineteen years.

The fact that global average temperatures have not significantly increased in the past sixteen to eighteen years is a difficult conundrum for climate researchers to explain. Such a long pause was not predicted by any of the climate computer models relied upon by the IPCC. The 2014 *Synthesis Report* deals with this awkward situation basically by waving it away, asserting that a sixteen-year pause is too short a time from which to draw any conclusions regarding future warming trends.

"Due to natural variability, trends based on short records are very sensitive to the beginning and end dates and do not in general reflect long-term climate trends," the IPCC's 2013 *Physical Science* report states. Nevertheless, the report asserts this: "The long-term climate model simulations show a trend in global-mean surface temperature from 1951 to 2012 that agrees with the observed trend (*very high confidence*)." What is the observed long-term trend? As previously noted, the trend has been increasing at the rate of 0.12°C per decade. To illustrate how sensitive temperature trends are to starting and ending dates, the IPCC report notes that for fifteen-year periods starting in 1995, 1996, and 1997 the global average temperature trends are 0.13, 0.14, and 0.07°C per decade, respectively. Nevertheless, the IPCC's 2014 *Synthesis Report* observes that the rate of warming over the past fifteen years has been only 0.05°C (0.09°F) per decade, which is considerably lower than the 0.12°C (0.22°F) per decade rate observed since 1951.

So are the models relied upon by the IPCC really all that good at simulating trends in global average temperature? An August 28, 2013, article in the journal *Nature Climate Change* suggests that they are not. In that study Canadian climate researchers pointed out that while global average temperatures rose over the past twenty years at a rate of 0.14 ± 0.06°C per decade, thirty-seven of the models relied upon by the IPCC simulated an average rise of 0.30 ± 0.02°C per decade. "The observed rate of warming given above is less than half of this simulated rate, and only a few simulations provide warming trends within the range of observational uncertainty," the authors conclude. It gets worse. For the period after 1998 until 2013, the researchers note, "The observed trend of 0.05 ± 0.08°C per decade is more than four times smaller than the average simulated trend of 0.21 ± 0.03°C per decade." The upshot is that, according to this study, the climate models are running four times hotter than the observed temperature trends.

John Christy, a climatologist at the University of Alabama in Huntsville who has long been skeptical of IPCC projections, compared the outputs of seventy-three climate models for the tropical troposphere used by the IPCC in its latest report with satellite and weather balloon temperature trends since 1979 until 2030. "The tropics is so important because that is where models show the clearest and most distinct signal of greenhouse warming—so that is where the comparison should be made (rather than, say, for temperatures in North Dakota)," explains Christy. "Plus, the key cloud and water vapor feedback processes occur in the tropics." When it comes to simulating the atmospheric temperature trends of the past thirty-five years, Christy found, all of the IPCC models are running hotter than the actual climate. To deflect accusations that he is cherry-picking data, Christy notes that his "comparisons start in 1979, so these are thirty-five-year time series comparisons"—rather longer than the fifteen-year periods whose importance the IPCC report downplays. The private research group Remote Sensing Systems (RSS) also analyzes the satellite temperature data. Like the University of Alabama climatologists RSS reports that "the troposphere has *not* (their emphasis) warmed as fast as almost all climate models predict." Comparing IPCC climate model simulations to actual temperatures, RSS finds that "after 1998, the observations are likely to be below the simulated values, indicating that the simulations as a whole are predicting too much warming." Even the IPCC report admits, "Most, though not all, of [the climate models] overestimate the observed warming trend in the tropical troposphere during the satellite period 1979–2012."

The IPCC *Physical Science* report also concedes that "[a]lmost all [climate model] historical simulations do not reproduce the observed recent warming hiatus." In fact, the IPCC's technical summary reveals that 111 out of 114 climate models run hotter than the actual observed temperature trend between 1998 and 2012. What accounts for this apparent predictive failure?

The IPCC's *Physical Science* report suggests with "medium confidence" that internal decadal variability is the cause of much of the difference between observations and the simulations. It's fair to say that this pause is something of an embarrassment to many in the climate research community, since the computer models failed to indicate that any such thing could happen. Spurred by the mismatch between computer projections and empirical data, lots of climate scientists have been trying to figure out why the average global temperature has not been increasing significantly.

For example, a 2010 study in *Science* attributed part of the temperature slowdown to decreases in stratospheric water vapor. A September 2011 article in *Nature Climate Change* outlined one of the more popular explanations for the pause. Specifically, the missing heat is supposedly hiding in the deep oceans.

A 2011 article in *Atmospheric Chemistry and Physics* suggested that a prolonged solar minimum combined with atmospheric aerosols left over from volcanic eruptions reduced the amount of heat reaching the surface of the planet.

But as noted previously, the most popular explanation by far for why the atmosphere was not warming even as greenhouse gas concentrations were rising was that the excess heat is hiding in the oceans. Some researchers in March 2014 argued in *Nature Climate Change* that the Pacific Ocean trade winds have speeded up, thus pushing extra greenhouse heat beneath the waves.

In August 2014 researchers at the University of Washington in Seattle and Ocean University of China in Qingdao countered in *Science* that the real reason the atmosphere is not warming is that changes in North Atlantic Ocean circulation are burying the extra warmth. The researchers reported that this process could go on for as long as another twenty years before the ocean begins releasing the stored heat, greatly boosting future rates of warming.

In late August 2014, a team of Japanese climatologists suggested in *Nature Climate Change* that natural variations in Pacific trade winds account for nearly half of the changes in temperature seen over the past three decades. The bad news, according the researchers, is that natural variation is now being overwhelmed by climate change caused by increasing greenhouse gas concentrations in the atmosphere.

Consequently, they predict that man-made warming will dominate future temperature trends soon and the hiatus will end.

Also in August, Australian researchers associated with the Pacific trade winds theory report in *Geophysical Research Letters* the results of thirty-one climate models. They find that "under high rates of greenhouse gas emissions there is little chance of a hiatus decade occurring beyond 2030, even in the event of a large volcanic eruption." As it happens, a study published in October 2014 in *Geophysical Research Letters* reports that the effects of volcanic particles on global atmospheric temperatures since 2000 have been underestimated, and they have actually cooled the planet by between 0.05 and 0.12 °C. A January 2015 study by researchers at the Lawrence Livermore National Laboratory in California published in *Geophysical Research Letters* bolstered the findings that sulfuric acid particles from small volcanic eruptions lingering in the boundary between the troposphere and the stratosphere reflect enough sunlight to contribute significantly to the warming hiatus. While volcanic particles in the atmosphere may be masking a warming trend, researchers from the National Center for Atmospheric Research and Duke University in their January 2015 study in *Climate Dynamics* concluded, "We do not find that aerosols exerted a significant global negative forcing over the last decade or so." In other words, man-made particulate air pollution has not been significantly cooling the planet during the hiatus. This finding contradicts earlier speculations that man-made aerosols might be responsible for a quarter of the discrepancy between model temperature projections and actual temperature trends.

In October 2014, two new papers in *Nature Climate Change* significantly challenged the popular theory that extra heat from greenhouse warming is being buried in the deep ocean. One study analyzed satellite and direct ocean temperature data from 2005 to 2013 and found the ocean abyss below 2,000 meters has not warmed measurably. Basically, satellite measurements between 2005 and 2013 find that sea level has been increasing at rate of 2.78 millimeters per year. Some 0.9 millimeters results from expansion due to warming, and 2.0 millimeters is due to additions of freshwater—for example, melting glaciers. Since 2.9 millimeters is greater than the measured increase of 2.78 millimeters, the researchers concluded that the deep ocean is likely cooling down and thus contracting. In its release describing the work of its researchers, the Jet Propulsion Laboratory noted, "The cold waters of Earth's deep ocean have not warmed measurably since 2005," thus "leaving unsolved the mystery of why global warming appears to have slowed in recent years."

In a simultaneously published companion paper, JPL researchers suggest that since the extra heat is not hiding in the deep oceans, it must be instead cached in the upper layers of the southern oceans, where it had gone unnoticed due to spotty measurements over the past thirty-five years. As the JPL release explained, "Using satellite measurements and climate simulations of sea level changes around the world, the new study found the global ocean absorbed far more heat in those 35 years than previously thought—a whopping 24 to 58 percent more than early estimates." It's always interesting when models find discrepancies in observational data.

The effort to explain or explain away the current warming hiatus is ongoing. In a January 2015 *Geophysical Research Letters* article a French research group noted, "The observed global mean surface air temperature (GMST) has not risen over the last 15 years, spurring outbreaks of skepticism regarding the nature of global warming and challenging the upper-range transient response of the current-generation global climate models." In an attempt to explain the hiatus, the researchers adjust their climate model to try to take into account how surface winds in the Pacific Ocean could be driving heat uptake in the ocean. Even so, they find that their model still overestimates warming compared to actual temperature trends.

Also in January, Duke University climatologists analyzed outputs from thirty-four of the climate models used by the Intergovernmental Panel on Climate Change (IPCC) in its Fifth Assessment Report. They report in the *Journal of Geophysical Research* that the models more or less tracked each other with regard to year-to-year temperature ups and downs but diverged in their explanations for decade-to-decade variability "such as why global mean surface temperatures warmed quickly during the 1980s and 1990s, but have remained relatively stable since then." Lead author Patrick Brown cautioned, "If you're worried about climate change in 2100, don't over-interpret short-term trends. Don't assume that the reduced rate of global warming over the last 10 years foreshadows what the climate will be like in 50 or 100 years." On the other hand, Brown also noted, "The inconsistencies we found among the models are a reality check showing we may not know as much as we thought we did."

In a January 2015 *Nature* article, two European climate researchers report the results of comparing the outputs of eighteen climate models used by the IPCC to simulate average global temperature trends from 1900 to 2010 to see how well they match with observed temperature trends. They find that the models actually do simulate similar lengthy hiatuses during that period; they just don't happen to coincide with the current observational hiatus. They find that due to natural variation, the observed warming might be at the upper or lower limit of simulated rates, but there is no indication of a systematic bias in model process. "Our conclusion is that climate models are fundamentally doing the right thing," University of Leeds researcher Piers Forster explained. "They [climate models] do in fact correctly represent these 15-year short-term fluctuations but because they are inherently chaotic they don't get them at the right time." The *Nature* article concludes, "The claim that climate models systematically overestimate the response to radiative forcing from increasing greenhouse gas concentrations therefore seems to be unfounded." Accordingly, the current pause in global average temperature increases is just the result of natural fluctuations in the climate and the man-made trend toward higher temperatures will resume eventually. What natural fluctuations might be responsible for slowing global temperature increases? In a February 2015 article in *Science* University of Pennsylvania climatologist Michael Mann and his colleagues used climate model simulations to estimate natural variability in North Atlantic and North Pacific Ocean temperatures. They conclude that temperatures in the northern Pacific just so happen to be in cold phase right now which has "produce[d] a slow-down or "false pause" in warming of the past decade."

Interestingly, the IPCC's *Synthesis Report* found that "ocean warming dominates the increase in energy stored in the climate system, accounting for more than 90% of the energy accumulated between 1971 and 2010 (*high confidence*) with only about 1% stored in the atmosphere." The report also suggested that most of the excess heat was stored in the upper ocean, but added that "it is *likely* that the ocean warmed from 700 m[eters] to 2000 m[eters] from 1957 to 2009 and from 3000 m[eters] to the bottom for the period 1992 to 2005." Of course, the JPL studies published the month before contradict this assertion.

Just how long the temperature pause must last before it would falsify the more catastrophic versions of man-made climate change obviously remains an open question for many researchers. For the time being, most are betting that it will get real hot real fast when the hiatus ends.

The upshot is that many researchers remain convinced that natural fluctuations in the climate unaccounted for in the computer models are responsible for keeping average global temperature flat for the past sixteen to eighteen years. The IPCC's *Physical Science* report asserts that the models cannot be expected to simulate the timing of the sort of natural climate variability that has produced the current sixteen- to eighteen-year pause. Georgia Tech climatologist Judith Curry contrarily observed, "If the IPCC attributes to the pause to natural internal variability, then this begs the question as to what extent the warming between 1975 and 2000 can also be explained by natural internal variability. Not to mention raising questions about the confidence that we should place in the IPCC's projections of future climate change."

Overall, the IPCC suggests that the difference between the models and the actual recent temperature trend "could be caused by some combination of (a) internal climate variability, (b) missing or incorrect radiative forcing, and (c) model response error." That is to say, the projections are off owing to pesky natural climate fluctuations, possible errors regarding estimates of how much warming a given increase in greenhouse gases will produce, and/or boosting temperature projections too high in response to given increases in greenhouse gases.

Nevertheless, the IPCC believes that the current temperature slow-down will soon end and states, "It is *more likely than not* that internal climate variability in the near-term will enhance and not counteract the surface warming expected to arise from the increasing anthropogenic forcing." In other words, when the warm-up resumes, it will soar. By how much? The IPCC *Physical Science* report projects, "The global mean surface temperature change for the period 2016–2035 relative to 1986–2005 will likely be in the range of 0.3°C to 0.7°C." This implies increases of 0.15°C to 0.35°C per decade. That would mean that warming could increase at nearly triple the rate the IPCC reports for the period after 1951, and seven times higher than the rate of increase it reports for the last fifteen years. Researchers from the Pacific Northwest National Laboratory published an article in *Nature Climate Change* in March 2015 in which they compared past rates of temperature change over forty-year periods with future projections. They predict that global average temperature will be increasing at a rate of 0.25C° per decade by 2020. That rate of change would be "unprecedented for at least the past 1,000 years." If average global temperature began to rise at this rate, it would vindicate the climate models. If not, then what?

What sorts of changes in "internal climate variability" might soon increase global average temperatures? Warmer and colder water sloshes back and forth periodically in the tropical Pacific Ocean, producing significant changes in global weather. When this El Niño Southern Oscillation (ENSO) pattern is in its warm phase, it substantially boosts the average global temperature. In 2014, many meteorologists were waiting to see if 2014–2015 would conjure up a big ENSO warm phase that would end the hiatus and finally increase global average temperatures above the big 1998 ENSO spike. As of January 2015, the National Oceanic and Atmospheric Administration noted the existence of mild El Niño–like conditions and suggested that Pacific Ocean sea surface temperatures have a good chance of subsiding to a neutral state in 2015.

The computer climate models are supposed to give policymakers reliable data regarding future trends in man-made global warming. The failure to predict the sixteen- to eighteen-year temperature hiatus has caused some policymakers to wonder if the findings in the IPCC's *Physical Science* report really do inspire the kind of confidence that could justify the entailed multitrillion-dollar bet on massive changes to humanity's energy supply programs.

The Science Is Settled—Climate Sensitivity

Another possible explanation for why the computer climate models may be running too hot is what the IPCC refers to as model response error. That is, they may overestimate the amount of warming that results from a given increase in greenhouse gas concentrations in the atmosphere. This brings up the crucial issue of climate sensitivity, conventionally defined as the amount of warming that doubling carbon dioxide in the atmosphere would eventually produce. Temperature increases lag increases in atmospheric carbon dioxide, so another important process is the transient climate response (TCR), which is the amount of warming expected at the time a carbon dioxide concentration crosses the doubling line.

In its 2007 report, the IPCC estimated that climate sensitivity was between 2° and 4.5°C, with the best estimate being 3°C. The 2013 IPCC *Physical Science* report drops the lower bound and finds that climate sensitivity is *likely* in the range 1.5° to 4.5°C. It also states that it is *extremely unlikely* climate sensitivity is less than 1°C and *very unlikely* to be greater than 6°C. In addition, the report notes with "high confidence" that the transient climate response is "likely in the range 1° to 2.5°C and extremely unlikely greater than 3°C, based on observed climate change and climate models." In IPCC parlance, *likely* means that the authors believe that there is more than a 66 percent chance that they've gotten the right estimate for climate sensitivity, whereas *extremely unlikely* means that they think there is less than

5 percent chance that they are wrong. Just how sensitive the climate is to increases in greenhouse gases is a controversial and hotly disputed area of climate research.

Several recent studies have reported that climate sensitivity could be lower than the *Physical Science* summary suggests. For example, an article in the June 2013 *Nature Geoscience* concluded that the "most likely value of equilibrium climate sensitivity based on the energy budget of the most recent decade is 2.0 °C, with a 5–95% confidence interval of 1.2–3.9 °C." A confidence interval is basically the probability that a value will fall between an upper and lower bound of a probability distribution. In other words, these researchers are 90 percent confident that climate sensitivity lies somewhere between 1.2° and 3.9°C. The researchers also reported that the best estimate for transient climate response based on observations of the most recent decade is 1.3°C, ranging between 0.9° to 2.0°C.

The *Nature Geoscience* estimate is a whole degree lower than the best estimate calculated by the IPCC in 2007. Interestingly, the 2013 *Physical Science* report, unlike its predecessor reports, provides no best estimate of climate sensitivity. Research on this topic continues. For example, in a March 2014 study, Norwegian researchers take into account the last ten years of temperature records and trends in ocean heat buildup. Their estimates for climate sensitivity and transient climate response are even lower. They report that the best estimate of climate sensitivity "is 1.8°C, with 90% C.I. [confidence interval], ranging from 0.9 to 3.2°C, which is tighter than most previously published estimates." They also calculate that there is only 1.4 percent chance that climate sensitivity would turn out to be greater than 4.5°C. Since the amount of warming that a doubling of carbon dioxide would produce is lower, so too is the transient climate response, which the researchers estimate to be about 1.4°C, ranging between 0.8° and 2.2°C.

Other researchers, however, come to more worrisome conclusions about how sensitive the climate is. Two new studies, one in March and another in May 2014, argue that the many researchers who have reported lower climate sensitivities have failed to take the cooling effects of various pollutants into account. Once the dampening of airborne particles like sulfates and black soot and ground-level ozone is properly included in calculations, the March *Nature Climate Change* study by GISS researcher Drew Shindell calculated a transient climate response of about 1.7°C. Building on Shindell's insights, researchers at Texas A&M University estimated in May 2014 that climate sensitivity is likely to be 3.0°C, ranging between 1.9° and 6.8°C. This range is a bit higher and wider than the one reported by the IPCC in 2013. However, more recent research cited earlier that now finds that man-made aerosols have had a negligible effect on current global average temperatures might suggest that these climate sensitivity ranges are too high.

In September 2014 the climate sensitivity research pendulum swung back toward a lower estimate. In their article published in *Climate Dynamics,* Georgia Tech climatologist Judith Curry and British statistician Nicholas Lewis reported results using temperature data from 1850 to 2011 along with estimates of the effects on climate of various factors taken from the IPCC's *Fifth Assessment Report,* such as land-use changes, volcanic activity, and atmospheric pollutants. They calculate that the best estimate for climate sensitivity is 1.64°C, with an uncertainty range of 1.05° to 4.05°C. The corresponding transient climate response is 1.33°C, with an uncertainty range of 0.90° to 2.50°C. These new values are at the low end of the IPCC climate sensitivity and transient climate response estimates.

Is there any way to resolve this scientific dispute? Yes. Wait and see what the climate actually does. But climatologists may have to wait at least a couple of decades before they can know the answer for sure. In an April 16, 2014, article in *Geophysical Research Letters,* researchers sort through various climate sensitivity scenarios ranging from a low of 1.5° to a high of 6.0°C. They calculate that it would take another twenty years of temperature observations for us to be confident that climate sensitivity is on the low end and more than fifty years of data to confirm the high end of the projections. This ongoing controversy is important because lower climate sensitivity would mean that future warming will be slower, giving humanity more time to adapt and to decarbonize its energy production technologies. Higher climate sensitivity would mean the opposite.

Ocean Acidification

As the oceans absorb carbon dioxide from the atmosphere, the amount of carbonic acid is increased, thus making the ocean more acidic. As noted previously, the acidity of the surface waters of the oceans has increased by about 26 percent since the beginning of the Industrial Revolution. The IPCC 2014 *Adaptation* report observes, "Impacts of ocean acidification range from changes in organismal physiology and behavior to population dynamics and will affect marine ecosystems for centuries if emissions continue." Some ocean denizens like plants and algae will most likely benefit from increased carbon dioxide levels, whereas other creatures such as corals and mollusks might suffer significant harm. Corals, echinoderms, and mollusks absorb carbonate minerals from the oceans to make their shells, and higher acid levels lower dissolved amounts of that mineral in seawater. Various computer model projections suggest that as acidity increases, it will be harder to calcifying creatures to survive. However, the 2014 *Adaptation* report observes, "Limits to adaptive capacity exist but remain largely unexplored."

Some researchers are worried that calcifiers like corals might reach a tipping point at which they "collapse." More recent experimental research on eight species of Pacific reef calcifiers has been a bit more reassuring. Biologists from California State University at Northridge report, "In contrast to previous studies that have predicted rapid decreases in calcification of corals and coral reefs exposed to [more than doubled carbon dioxide], our study, performed at the organismic level on eight of the main calcifiers in Moorea, suggests that tropical reefs might not be affected by OA [ocean acidification] as strongly or as rapidly as previously supposed." They found that overall reef calcification declined by about 10 percent when carbon dioxide was doubled. In addition, while calcification rates declined with higher and higher levels of carbon dioxide, even when carbon dioxide reached ten times its preindustrial levels, the researchers could identify no tipping points beyond which calcification collapsed. Studies on cold-water Mediterranean corals similarly found that their rates of calcification remained constant even when exposed to levels of carbon dioxide in high-end projections for the end of this century. Nevertheless, published reviews of research on the effects of ocean acidification resulting from high levels of extra carbon dioxide find the overall effects on marine organisms are negative. Of course, if emissions are cut to keep future temperature increases down, that would also limit the effects of ocean acidification.

How Much Will Global Warming Cost?

Assume global warming. There are two ways to address concerns about warming: adaptation and mitigation. In 2014, the IPCC issued two reports dealing with both sorts of responses. *Climate Change 2014: Impacts, Adaptation, and Vulnerability* (hereafter the *Adaptation* report) describes adaptation as the "process of adjustment to actual or expected climate and its effects." *Climate Change 2014: Mitigation* (hereafter the *Mitigation* report) defines mitigation as "a human intervention to reduce the sources or enhance the sinks of greenhouse gases." Mitigation basically means cutting the emissions of greenhouse gases like carbon dioxide and/or figuring out how to suck carbon dioxide out of the atmosphere—for example, planting more forests. In other words, people can take steps to defend themselves against the impacts of man-made climate change and/or reduce impacts by trying to slow or stop man-made climate change. Most climate researchers believe that some additional warming is inevitable, so people will have to engage in both activities.

The IPCC reports offer cost estimates for both adaptation and mitigation. The 2014 *Adaptation* report reckons, assuming that the world takes no steps to deal with climate change, that "global annual economic losses for additional temperature increases of around 2°C are between 0.2 and 2.0 percent of income." The report adds, "Losses are more likely than not to be greater, rather than smaller, than this range."

In a 2010 *Proceedings of the National Academy of Sciences* article, Yale economist William Nordhaus assumed that humanity does nothing to cut greenhouse gas emissions. Nordhaus uses an integrated assessment model that combines the scientific and socioeconomic aspects of climate change to assess policy options for climate change control. His RICE-2010 integrated assessment model found that "of the estimated damages in the uncontrolled (baseline) case, those damages in 2095 are $12 trillion, or 2.8% of global output, for a global temperature increase of 3.4°C above 1900 levels." Nordhaus's estimate evidently assumes that the world's economy will grow at about 2.5 percent annually, reaching a total GDP of roughly $450 trillion in 2095.

What might the world's economy look like by 2100 if no policies are adopted with the aim of mitigating or adapting to climate change? In 2012, the IPCC asked the economists in the Environment Directorate at the Organisation for Economic Co-operation and Development to peer into the future and devise a plausible set of shared socioeconomic pathways (SSPs) to the year 2100. The OECD economists came up with five baseline scenarios. Let's take a look at a couple of the scenarios to get some idea of how the world's economy might evolve over the remainder of this century. The OECD analysis begins in 2010 with a world population of 6.8 billion and a total world gross product of $67 trillion (2005 dollars), yielding a global per capita income just shy of $10,000. For reference the OECD notes that US 2010 per capita income averaged $42,000.

The SSP2 scenario is described as the "middle of the road" projection in which "trends typical of recent decades continue, with some progress towards achieving development goals, reductions in resource and energy intensity at historic rates, and slowly decreasing fossil fuel dependency." If economic and demographic history unfolds as that scenario suggests, world population will have peaked at around 9.6 billion in 2065 and fallen to just over 9 billion by 2100. The world's economy will have grown more than eightfold, from $67 trillion to $577 trillion (2005 dollars). Average income per person globally will have increased from around $10,000 today to $60,000 by 2100. US annual incomes would average just over $100,000.

In the SSP5 "conventional development" scenario, the world economy grows flat out, which "leads to an energy system dominated by fossil fuels, resulting in high GHG emissions and challenges to mitigation." Because there is more urbanization and because there are higher levels of education, world population peaks at 8.6 billion in 2055 and will have fallen to 7.4 billion by 2100. The world's economy will grow fifteenfold to just over $1 quadrillion, and the average person in 2100 will be earning about $138,000 per year. US annual incomes would exceed $187,000 per capita.

It is of more than passing interest that people living in the warmer world of SSP5 are much better off than people in the cooler SSP2 world. The OECD analysis adds with regard to climate change in this scenario that the much richer and more highly educated people in 2100 will face "lower socio-environmental challenges to adaptation result[ing] from attainment of human development goals, robust economic growth, highly engineered infrastructure with redundancy to minimize disruptions from extreme events, and highly managed ecosystems." In other words, greater wealth and advanced technologies will significantly enhance the capabilities of people to deal with whatever the deleterious consequences of climate change turn out to be.

As noted above, the IPCC estimates that failure to adapt to climate change will reduce future incomes by between 0.2 to 2 percent for temperatures exceeding 2°C. Yale's William Nordhaus is one of the more accomplished researchers in this area, trying to calculate the costs and benefits of climate change. In his 2013 book *The Climate Casino: Risk, Uncertainty, and Economics for a Warming World,* Nordhaus notes that a survey of studies that try to estimate the aggregated damages that climate change might inflict at 2.5°C warming comes in at an average of about 1.5 percent of global output. The highest climate damage estimate Nordhaus cites is a 5 percent reduction in income. The much criticized 2006 *Stern Review: The Economics of Climate Change* suggested that the business-as-usual path of economic growth and greenhouse gas emissions could even reduce future incomes by as much as 20 percent.

Future temperatures will perhaps exceed these, but transient climate response temperatures over the remainder of the century are likely to be close to the 2.5°C benchmark cited by the IPCC. In the scenarios sketched out above, a 2 percent loss of income would mean that the $60,000 and $138,000 per capita income averages would fall to $58,800 and $135,240, respectively. Stern's more apocalyptic estimate would cut 2100 per capita incomes to $48,000 and $110,400, respectively. How much should people living now on incomes averaging $10,000 per year spend to make sure that people whose incomes will likely be 6 to 14 times higher aren't reduced by a couple of percentage points? As Nordhaus observes, "Most philosophers and economists hold that rich generations have a lower ethical claim on resources than poorer generations."

The Costs and Benefits of Trying to Stop Warming

Making the heroic set of assumptions that all countries of the world begin mitigation immediately and adopt a single global carbon price, and all key low- and no-carbon technologies are now available, the IPCC's 2014 *Mitigation* report estimates that keeping carbon dioxide concentrations below 450 ppm by 2100 would result in "an annualized reduction of consumption growth by 0.04 to 0.14 percentage points over the century relative to annualized consumption growth in the baseline that is between 1.6 percent and 3 percent per year." The median estimate in reduced annualized growth in consumption is 0.06 percent.

The IPCC *Mitigation* report notes that the optimal scenario that it sketches out for keeping greenhouse gas concentrations below 450 ppm would cut future incomes by 2100 by between 3 and 11 percent. How much would that be? As was done with regard to the losses from a lack of adaptation, let's look at how much the worst-case mitigation scenario might reduce future incomes. Without extra mitigation, the increase of global gross product to $577 trillion in the middle-of-the-road scenario implies an economic growth rate of 2.42 percent between 2010 and 2100. Cutting that growth rate by 0.14 percentage point to 2.28 percent yields an income of $510 trillion in 2100, reducing per capita incomes from $60,000 to $57,000 per capita. Growth in the conventional-development scenario is cut from an implied 3.07 percent to 2.93 percent, reducing overall income from over $1.015 quadrillion to $901 trillion and cutting average incomes from $138,000 to $122,000.

All of these figures must be taken with a vat of salt since they are projections for economic, demographic, and biophysical events nearly a century from now. That being acknowledged, projected IPCC income losses that would result from doing nothing to adapt to climate change appear to be roughly comparable to the losses in income that would occur following efforts to slow climate change. In other words, it appears that doing nothing about climate change now will cost future generations about the same as doing something now.

Climate Change Is Not Increasing Damage—Yet

People concerned about catastrophic man-made climate change have been seeking evidence that it is boosting risks now among the weather damage and loss data. "This is climate change. We were warned about extreme weather. Not just hot weather, but extreme weather," Senator Barbara Boxer (D-CA) declared on the floor of the US Senate in May 2013 in response to the tornado that devastated Moore, Oklahoma. "You're going to have tornadoes and all the rest." The activists over at Greenpeace did not put too fine a point on the destruction caused by Superstorm Sandy hitting New York and New Jersey in the fall of 2013: "Hurricane Sandy = Climate Change." One problem: Researchers can find no such trends with respect to the damage caused by tornadoes and hurricanes.

The 2012 United Nations' *Special Report for Managing the Risks of Extreme Events and Disasters to Advance Climate Change Adaptation* (SREX) projects man-made global warming will boost the damage caused by heat waves, coastal floods, and droughts as they get worse by the end of the century. The researchers, however, could not draw firm conclusions about the effects of climate change on any current trends in hurricanes, typhoons, hailstorms, or tornadoes. The IPCC's *Climate Change 2014: Synthesis Report* noted that there is low confidence that climate change has so far affected any global trends toward increased flooding, hurricanes and typhoons, or droughts. That report did note that "increasing exposure of people and economic assets has been the major cause of long-term increases in economic losses from weather- and climate-related disasters." In other words, the weather is not necessarily getting worse; there are simply more people and property for storms to damage.

The SREX study expressed medium confidence that droughts had increased in some areas as a result of man-made climate change. An authoritative November 2012 article in *Nature* later found that the previously reported increase in global drought is overestimated because the widely used Palmer Drought Severity Index doesn't, among other issues, properly account for evaporation rates. Consequently, based on more realistic calculations that take into account changes in available energy, humidity and wind speed, the researchers concluded that "there has been little change in drought over the past 60 years." The 2013 IPCC *Physical Science* report essentially concurred, observing that there was "*low confidence* in detection and attribution of changes in drought over global land areas." (In December 2014, NOAA issued a study that concluded that man-made climate change was not a factor in the extreme drought that has beset California since 2011.)

The 2012 SREX report acknowledges, "In many regions, the main drivers for future increases in economic losses due to some climate extremes will be socioeconomic in nature." In other words, any surge in weather disaster damage is largely due to an increase in what can potentially be destroyed and the number of people exposed to it.

Can researchers now discern any effect that the recent increase in global average temperature has had on people and their property? Not really.

For example, a 2011 report by the libertarian Reason Foundation, *Wealth and Safety: The Amazing Decline in Deaths from Extreme Weather in an Era of Global Warming, 1900–2010*, notes, "Aggregate mortality attributed to all extreme weather events globally has declined by more than 90 percent since the 1920s, in spite of a four-fold rise in population and much more complete reporting of such events." The death rate from droughts is 99.9 percent lower than it was in the 1920s; the death rate from floods is 98 percent lower; and the death rate from big storms like hurricanes has declined more than 55 percent since the 1970s.

Keep in mind that the death rate due to extreme weather between 2001 and 2010 averaged about 38,000 per year compared to about 59 million annual deaths for all causes. The Reason Foundation report concludes, "While extreme weather-related events, because of their episodic nature, garner plenty of attention worldwide, their contribution to the global mortality burden—0.07 percent of global deaths—is relatively minor."

What about economic losses? The IPCC's 2014 *Adaptation* report observes, "Economic costs of extreme weather events have increased over the period 1960–2000, with insured losses increasing more rapidly than overall losses." The report, however, goes on to note that "the greatest contributor to increased cost is rising exposure associated with population growth and growing value of assets." Similarly, the 2014 *Synthesis* report notes that "increasing exposure of people and economic assets has been the major cause of long-term increases in economic losses from weather- and climate-related disasters." To repeat: There is more damage because there are more people and more stuff to be harmed by storms, floods, wind, and drought.

In order to see if climate change is adding to the destruction, researchers "normalize" losses by taking into account the number of people and the value of the property exposed to extreme weather events. For example, far more people live in Florida now than fifty years ago, with lots more houses and businesses, so hurricanes that strike there today are more likely to cause damage than those than hit that state, say, in the 1920s. Taking such increases in population and wealth into account, the IPCC *Adaptation* report contradicts the claims of Senator Boxer and Greenpeace. "Studies of normalized losses from extreme winds associated with hurricanes in the U.S. and the Caribbean, tornadoes in the U.S. and wind storms in Europe have failed to detect trends consistent with anthropogenic climate change," notes the report.

A January 2011 review article, "Have Disaster Losses Increased Due to Anthropogenic Climate Change?," by Laurens Bouwer, which was published in the *Bulletin of the American Meteorological Society* (BAMS), surveyed twenty-two studies looking at trends in natural hazard losses. Bouwer, who was then a researcher at the Institute for Environmental Studies at Vrije University in the Netherlands, included studies that all looked at economic losses, covered at least thirty years of data, and were peer reviewed. The BAMS review found, "The studies show no trends in the losses, corrected for change (increases) in population and capital at risk, that could be attributed to anthropogenic climate change. Therefore, it can be concluded that anthropogenic climate change so far has not had a significant impact on losses from natural disasters."

Another 2010 study, "Normalizing Economic Loss from Natural Disasters: Global Analysis," by Eric Neumayer and Fabian Barthel, two researchers associated with the Grantham Research Institute on Climate Change and the Environment at the London School of Economics and Political Science, also probed trends in weather disaster loss data in search of a global warming signal. Besides using conventional techniques that take into account increases in population and wealth to normalize losses, they also develop an alternative technique that looks at relative losses over time. Briefly, their new measure looks at how much actual loss occurred relative to the amount that was at risk. For example, what percentage of wealth in Miami was destroyed by hurricanes in 1920 versus 2010? If the actual-to-potential-loss ratio is increasing over time, this suggests that the weather is having a growing impact.

Analyzing weather disasters between 1980 and 2009, Neumayer

and Barthel find, "Both methods lead to the same result for all disasters: no significant trend over time according to the conventional method, a marginally significant downward trend according to the alternative method." Applying both normalization methods, they find no significant trends in weather related losses for both developed and developing countries. Looking regionally at North America, Western Europe, Latin America and the Caribbean, and South and East Asia also uncovers no statistically significant trend in losses caused by weather disasters. In addition, two 2009 studies found no upward trend in normalized losses dues to windstorms or floods in Western Europe since 1970. One concluded, "Results show no detectable sign of human-induced climate change in normalised flood losses in Europe."

Neumayer and Barthel, using their alternative normalization method, do identify a "strongly negative trend" in normalized weather disaster damages in developed countries. Translation: The amount of damage caused by severe weather is declining in relative terms. They speculate, "This could possibly indicate a stronger capability of richer nations to fund defensive mitigating measures, which decrease vulnerability to natural disasters over time." Richer societies are likely reducing their weather losses by establishing better early warning systems, enacting stronger building codes, and constructing firmer levees. People may be protecting themselves ever better against the consequences of storms and floods, even though the weather is getting worse.

A review of the relevant weather damage studies, according to the IPCC *Adaptation* report, concluded, "A majority of studies have found no detectable trend in normalized losses." Given that the data it reviewed showed no significant evidence for current damages attributable to climate change, it is odd that the 2014 *Adaptation* report did nevertheless feebly conclude that "climate change cannot be excluded as at least one of the drivers involved in changes of normalized losses over time in some regions and for some hazards."

It is generally agreed that the average temperatures over land have increased by about 0.7°C since the 1950s. Looking toward the end of the twenty-first century, the 2012 SREX report relies on computer model projections, which suggest that one-in-twenty-year hottest day events are to become one-in-two-year events. The report also projects that inundations that once happened every twenty years are likely to occur every five years.

Sounds bad, but that's nearly a hundred years from now. With regard to the next few decades, the SREX researchers more sanguinely report: "Projected changes in climate extremes under different emissions scenarios generally do not strongly diverge in the coming two to three decades, but these signals are relatively small compared to natural climate variability over this time frame. Even the sign of projected changes in some climate extremes over this time frame is uncertain." That means that weather extremes for the next several decades will likely be within the bounds of natural variation, making it almost impossible to discern any effect of man-made climate change on them. In other words, whatever weather disasters do occur will not be on a scale or frequency beyond those that humanity has experienced in recent decades.

Although no upward trend in weather damages can be found in developing countries, the UN's SREX report does note that fatality rates and economic losses as a proportion of GDP from weather disasters are higher in poor countries. In fact, between 1970 and 2008, 95 percent of deaths from natural disasters occurred in developing countries. Bad weather produces death and destruction largely when it encounters poverty.

This insight prompts two observations: First, recent research indicates that man-made climate change has not been nor is it likely to be a big contributor to losses stemming from weather disasters in the next few decades. Second, boosting the wealth of poor people through economic growth is their best protection against meteorological disasters in the long run, whether fueled by future man-made climate change or not.

The Abject Failure of UN Climate Change Negotiations

One thing is for sure, the current international negotiating process under the auspices of the United Nations Framework Convention on Climate Change (UNFCCC) has been a near complete failure. The original goal of this creaky process set in motion back at the 1992 Earth Summit in Rio de Janeiro was to cut the greenhouse gas emissions of rich developed countries by 2012 to about 5 percent below the levels they emitted in 1990. A cap-and-trade mechanism was eventually devised under the Kyoto Protocol in 1997 and came into effect in 2005. The general idea was that all of the rich countries would set a cap on how much carbon dioxide could be emitted and issue permits for amounts under that cap. The permits would be tradable so that companies able to more cheaply cut their emissions could sell their excess permits to companies where reductions would be more costly. The cap would decline over time, and this carbon market would set a price on each ton of carbon dioxide. Setting a price on carbon dioxide was supposed to encourage the development of innovative low-carbon energy technologies that would replace fossil fuels as the main source of power.

It hasn't worked out that way. First, the United States, then the largest emitter of carbon dioxide, refused to adopt the Kyoto Protocol. In fact, the US Senate passed a resolution in 1997 by a vote of 95 to 0 that it was not the sense of the Senate that the United States should sign the Kyoto Protocol. The treaty was never submitted to the Senate for ratification. Some signatories, including Canada and Japan, ignored their Kyoto Protocol obligations and burned even more fossil fuels. In 2012, Canada, Japan, and Russia refused to make any further commitments under the Kyoto Protocol. Only the European Union actually established a cap-and-trade carbon market that has suffered periodic price collapses.

Nevertheless, the UNFCCC process is supposed to yield a new treaty at its 2015 meeting in Paris that obligates countries to begin controlling their greenhouse gas emissions by 2020. The IPCC's 2014 *Mitigation* report estimates that to have a good chance of keeping the global average temperature from rising above 2°C, global greenhouse gas emissions by 2050 must be between 40 and 70 percent lower than they were in 2010. Greenhouse gas emissions will have to be entirely eliminated by 2100.

So where do we stand now with respect to future warming? Since the failure of the UNFCCC climate change negotiations to adopt a binding treaty at the Copenhagen conference in 2009, many countries have made a variety of nonbinding "pledges" to cut back on their greenhouse gas emissions. For example, the United States under the Obama administration has promised to reduce emissions to below 17 percent of their 2005 levels by 2020. In November 2014, President Obama further pledged to reduce US emissions by 26 to 28 percent by 2025. Adding all of these pledges up, IPCC's 2014 *Mitigation* report suggests that they are "broadly consistent" with scenarios that would "keep temperature change below 3°C relative to preindustrial levels."

A Great Leap Forward on Climate Change?

At the Asia-Pacific Economic Cooperation summit in November 2014, US president Barack Obama and Chinese president Xi Jinping issued a "joint announcement on climate change" in which each country made pledges about how they intend to handle future emissions of their greenhouse gases. The announcement was hailed by most environmental groups and much of the media as "historic," a "breakthrough," and a "game changer." Careful parsing of the text's diplomatic jargon suggests that the joint announcement is in fact likely to be none of those.

To understand the nebulous nature of the announcement, don't focus first on the promised trajectories of future greenhouse gas emissions by both countries. Instead consider the loopholes. For example, this bit of climate change diplomatic arcana in which the two countries promise to work together "to adopt a protocol, another legal instrument or an agreed outcome with legal force under the Convention applicable to all Parties at the United Nations Climate Conference in Paris in 2015."

That convoluted phraseology was hammered out at as a compromise at the 2011 Durban climate conference. The European Union was then strongly insisting that the UN climate conferees commit to "a protocol or other legal instrument" as the ultimate goal for a comprehensive global treaty in 2015. Why? Because that exact language had earlier propelled the agreement to the Kyoto Protocol, which established the only legally binding emissions reduction targets on any countries.

China and India, however, objected and sought to water down the language by including "or an agreed outcome with legal force." The Chinese and Indians evidently believe that that phraseology suggests whatever climate negotiations do achieve by 2015, the result will be that they still will have fewer obligations to reduce their greenhouse gas emissions than will rich developing countries.

But what about the phrase "applicable to all Parties"? At Durban, the United States insisted that in any future climate agreement, "legal parity" must apply to big emerging economies like China, India, and Brazil. That means that they would be bound to cut their emissions in the same way that industrialized countries are. If China, India, and Brazil will not accept legally binding targets, then neither would the United States.

The joint announcement, most likely at the insistence of China, also reaffirmed "the principle of common but differentiated responsibilities and respective capabilities, in light of different national circumstances" enshrined in the United Nations Framework Convention on Climate Change. China has consistently interpreted that principle as meaning that countries that were rich and developed in 1992 when the convention was adopted are obligated to cut their emissions, while countries that were then poor are not.

What about the actual emissions pledges? The joint announcement states that the United States intends to achieve an economy-wide target of reducing its emissions by 26 to 28 percent below its 2005 level in 2025 and to make best efforts to reduce its emissions by 28 percent. Additionally, China intends to achieve the peaking of CO_2 emissions around 2030 and to make best efforts to peak early and intends to increase the share of nonfossil fuels in primary energy consumption to around 20 percent by 2030. The crucial word here is "intends." It is clear that the announcement is not meant to create any new obligations.

While China declared that its carbon dioxide emissions (not greenhouse gases) will peak by 2030, the announcement said nothing about the level at which they will peak. So at what level might China's emissions peak? Assuming the recent 3 percent annual increase in China's carbon dioxide emissions continues for the next sixteen years, they would reach 16 gigatonnes by 2030.

In 2005, the United States emitted the equivalent of 7.26 gigatonnes of carbon dioxide. So cutting emissions by 28 percent by 2025 implies emissions of 5.23 gigatonnes in 2025, which is about the amount that the United States emitted in 1992. Assuming that Chinese emissions did peak in 2030, the country could by then be emitting three times more than the United States.

Looking at the previously announced energy and climate policies of both the United States and China, the new pledges appear to add little to their existing plans to reduce their emissions. The new Obama pledges basically track the reductions that would result from the administration's plan to boost automobile fuel economy standards to 54.5 miles per gallon by 2025 and the Environmental Protection Agency's new scheme to cut by 2030 the carbon dioxide emissions from electric power plants by 30 percent below their 2005 level. Xi was no doubt aware that a week earlier an analysis of demographic, urbanization, and industrial trends by the Chinese Academy of Social Sciences had predicted that China's emissions peak would occur between 2025 and 2040.

Ultimately, there is a critical mismatch between the two countries' pledges. The United States undertakes to make actual cuts in its emissions over the next decade while China promises that it will do so beginning in sixteen years. Supporters hope that the joint announcement is the prelude to a "great leap forward" to a broad and binding global climate change agreement at Paris in 2015. Perhaps, but the United States and China left themselves plenty of room to step back if their pledges become inconvenient.

The Rocky Road to the Paris Climate Talks

At the December 2015 UN climate change conference in Paris, the nations of the world are supposed to adopt some kind of legally binding agreement to comprehensively address the problem of man-made climate change. That goal may be out of reach. Why? Because the interests of the rich and poor countries just don't converge on how to handle climate change, as the latest UN conference in Lima, Peru, made very clear in December 2014. Poor countries are demanding hundreds of billions in climate aid upfront before they agree to give up cheap fossil fuels. And rich countries won't hand over the cash unless poor countries first make credible commitments to cut their greenhouse gas emissions.

In early 2015, each nation is supposed to register publicly its voluntary intended nationally determined contributions (INDCs) to how it plans to address climate change. Attached to the *Lima Call for Climate Action* is a preliminary draft document outlining various options for a Paris agreement. This is a Chinese menu of provisions that highlights just how much discord there is over global climate policy. For example, the draft offers several options with regard to setting a firm goal for greenhouse gas emissions cuts. Countries might agree to cut emissions to 40 to 70 percent below their 2010 levels by 2050; or cut them by 50 percent below their 1990s levels with a continued decline thereafter; or go for full decarbonization by 2050. Or rich countries could agree that their emissions will peak in 2015 and then aim for zero net emissions by 2050.

The section on the financial resources to be provided to poor countries to help them to adapt to climate change and to pay for losses stemming from climate change suggests an annual floor of $100 billion in aid from rich countries; or, alternately, the agreement might not specify any amount of climate aid at all. Under the proposed provisions dealing with sources of finance, one option states that climate aid should primarily come from the government budgets of the rich countries. In a second option, private funds would play a greater role. Also undecided is whether countries will have the right to assess and challenge data issued by other countries with regard to their treaty commitments. The climate negotiators in Paris will also need to figure out whether the parties will have to update their INDCs every five years or every ten years.

Carbon Market Follies

Unfortunately, as noted earlier, the current model for controlling the global emissions of greenhouse gases like carbon dioxide is a cap-and-trade scheme devised under the Kyoto Protocol. To comply with its obligations under the Kyoto Protocol, the European Union implemented its Emissions Trading Scheme (ETS) back in 2005. The ETS covers the output of about 12,000 big emitters, whose CO_2 amounts to roughly half of the European Union's total emissions.

Under cap-and-trade schemes, governments set a limit on how much of a pollutant—in the case of man-made global warming, chiefly carbon dioxide—utilities and other enterprises can emit and then allocate permits to them. The permits can then be bought and sold on an open market. Manufacturers, for example, who can cheaply abate their emissions will have some permits left over. The cheap abaters can sell their extra permits to other enterprises that find it more expensive to reduce their emissions. In this way, a market in pollution permits is supposed to find the cheapest way to cut emissions.

That is the ideal, but implementing the ETS has been far from ideal. For example, in May 2006, an audit showed that several EU governments had issued permits for 66 million tons *more* CO_2 than was actually being emitted. Traders immediately realized that the supply of permits was not scarce, so the price of carbon dioxide allowances promptly collapsed to less than 9 euros per ton. By February 2007, an allowance to emit a ton of CO_2 could be had for less than a euro. European governments later tightened limits on carbon dioxide emissions and permit prices recovered in the second trading period until the advent of the financial crisis in 2008 forced a dramatic economic slowdown.

The steep decline in economic activity has lowered CO_2 emissions, producing a surplus of carbon permits among companies in the EU's emissions trading scheme. Consequently, by April 2014, the prices of carbon permits had fallen to a record low 2.46 euros per ton. Due to tinkering by European lawmakers, permit prices had risen to around 7 euros in January 2015. This is far under the price of 25 euros per ton that most analysts believe is necessary to drive energy producers to seek lower carbon sources of power.

The main point is that such price volatility means that companies have great difficulty in planning their infrastructure investments. There is very little evidence that the ETS has driven large-scale capital investments in energy production aimed at reducing the emissions of greenhouse gases among firms and facilities subject to the system. If carbon dioxide trading does not induce those kinds of investments, then it clearly has failed.

Windfall Profits for Corporations— Higher Prices for Consumers

In addition to being ineffective at encouraging investment in low-carbon energy technologies, permits have been distributed in such a way that they have provided billions of euros in windfall profits to polluters. How does this work? Beginning in 2005, the ETS cap-and-trade scheme handed out nearly all of its emissions permits gratis. Hold on, you might say: If the emitters are getting permits for free, why don't they pass along the lower costs to their customers?

Think of it in terms of an analogy put forward by left-leaning economists James Barrett and Kristen Sheeran: Tickets from scalpers for the last World Cup Soccer championship games were going for more than 200 euros, about double their face value. Would the price have been lower if a scalper had found them on the ground? No. "The supply and demand for tickets is the same no matter how much the scalper paid for them, and so the price he charges you will also be the same no matter how he got them," note Sheeran and Barrett. Or think of it this way, if someone gave you a bundle of cash worth a thousand euros, you would not be inclined to sell them to another person for less than a thousand, would you? The same thing is true of carbon dioxide emissions permits.

Giving away permits for free to industry is largely equivalent to a carbon tax in which the tax revenues are given to energy company stockholders, not spent on behalf of consumers and taxpayers. Before the carbon market's initial collapse in April 2006, the consultancy IPA Energy estimated that permits granted to British and German utilities fattened their bottom lines by 1 billion euros and 6 to 8 billion euros, respectively. And British and German consumers paid more for their electricity on top of that.

One way to correct the most egregious flaws in current cap-and-trade schemes would be to adopt cap-and-auction instead. Auctioning permits is very much like imposing a carbon tax. In this case, the government sets an overall emissions limit and emitters have to buy allowances from the government every year. The chief difference between a cap-and-auction scheme and a carbon tax is that the price of the allowances will vary from year to year. Once again, this variability in permit prices introduces uncertainty in the infrastructure planning of firms.

A 2011 report by the Swiss bank UBS concluded that the Emissions Trading Scheme will cost European consumers $277 billion for "almost zero impact." If the European Union, which has a relatively robust governmental institutions and the rule of law, couldn't effectively implement a cap-and-trade carbon market, there is no chance that the entire world encompassing China, Russia, India, Nigeria, Brazil, Iran, Mexico, Indonesia, and Saudi Arabia can do so.

Why Not a Carbon Tax?

Many economists think that a better option for rationing carbon would be a gradually rising tax on fuels that emit carbon dioxide. As the tax increases, industries and consumers would cut back on their use of more expensive fossil fuel energy and switch to using energy produced by low-carbon and no-carbon technologies. This process would lead to lower carbon dioxide emissions over time.

For instance, economists such as Harvard University's Gregory Mankiw and Yale University's William Nordhaus advocate imposing a tax on all kinds of carbon-based fuels at the wholesale stage, at the point where they emerge from under the ground. Thus, utilities and refiners who take raw coal, oil, and natural gas as inputs would pay a tax on these fuels. The extra cost would get passed downstream to all subsequent consumers. Thus carbon taxes would encourage conservation and low-carbon energy innovation. Since the tax is levied on how much carbon a fuel contains, it would make fuels like coal less attractive compared with low-carbon fuels like natural gas or even renewable energy like solar and wind power. Ideally, carbon tax revenues would be used to cut domestic taxes such as the payroll tax or the individual income tax, thus offsetting some of the pain of higher energy prices.

Internationally, one of the big advantages of a carbon tax is that it avoids the baseline quandary that bedevils carbon markets. For example, signatories to the Kyoto Protocol are supposed to cut their emissions of greenhouse gases by 5 percent below what they emitted in 1990. Why? That goal has no relationship to any specific environmental policy objective. In fact, achieving the cuts specified by the Kyoto Protocol goals would reduce projected average global temperatures by only a minuscule 0.07°C by 2050.

As the now-moribund international negotiations about what to do after the Kyoto Protocol show, it is very difficult to get many countries to agree to new global emissions baselines. Also, where should baselines be established for rapidly growing economies like China, India, and Brazil, whose energy use and emissions are expected to more than double by 2030? Under the Kyoto Protocol, the natural baseline is what emissions would be without any restraints. However, calculating or predicting what a country's emissions will be twenty to thirty years in the future is impossible to do with accuracy.

Under a pollution tax scheme, argues Yale economist William Nordhaus, "The natural baseline is a zero-carbon-tax level of emissions, which is a straightforward calculation for old and new countries. Countries' efforts are then judged relative to that baseline."

Another advantage is that the tax could be phased in as the average incomes of poor countries reach a certain threshold. For example, carbon taxes might start to kick in when national income reaches

$10,000 per capita, which is slightly higher than China's current level. More generally, having a defined tax rate makes it easy for firms in developed and developing economies alike to predict the future impact of climate policy on their bottom line—something that is considerably harder to do when the government is handing out permits every year.

A tax avoids the messy and contentious process of allocating allowances to countries internationally and among companies domestically. For example, nations could negotiate a much more transparent treaty than the Kyoto Protocol and establish a system of globally harmonized domestic carbon taxes. Harmonized taxes offer relative price stability, and taxes on carbon emissions can be raised gradually and predictably over time so that governments, industries, and consumers can all see what the price of carbon-based fuels will be over future decades and can make investment and purchase decisions accordingly.

Nordhaus further argues that carbon markets are "much more susceptible to corruption" than are tax schemes. "An emissions-trading system creates valuable tradable assets in the form of tradable emissions permits and allocates these to different countries," writes Nordhaus. "Limiting emissions creates a scarcity where none previously existed and in essence prints money for those in control of the permits."

So a carbon tax offers less opportunity for corruption because it does not create artificial scarcities and monopolies. Of course, governments can engage in chicanery by dispensing tax breaks and subsidies to favored companies and industries. But Nordhaus analogizes carbon allowances to quotas in international trade and carbon taxes to tariffs: overall, it's been a lot easier to manage tariffs than quotas.

Save the Climate: Cut Subsidies

The first rule for getting out of a hole is to stop digging. In this case, it's crazy to pay people to burn more fossil fuels if one is concerned about man-made global warming. The International Energy Agency estimates that government consumption subsidies for fossil fuels amounted to $544 billion in 2012. Ending subsidies would encourage consumers and producers to cut back on the use of fossil fuels, which in turn would reduce carbon dioxide emissions. And that would save taxpayers a great deal of money.

Nitrous oxide exists naturally in the atmosphere, but as a result of human activities, its concentration has increased by 20 percent over pre-industrial levels, making it the third most important greenhouse gas after carbon dioxide and methane. Nitrous oxide is a long-lived gas that has a global warming potential of 310, meaning one molecule traps over 310 times more heat than a molecule of carbon dioxide. The amount of nitrous oxide put into the atmosphere is equivalent to about 3 giga-tonnes of carbon dioxide, which approximates the emissions of half of the world's entire vehicle fleet. In addition, nitrous oxide depletes the stratospheric ozone layer that shields the earth's surface from damaging ultraviolet light. So reducing nitrous oxide emissions is a twofer—cutting it lowers the temperature and protects the ozone layer.

Two-thirds of human nitrous oxide emissions come from agricultural activities—for example, using nitrogen fertilizer or livestock waste management. It is not an exaggeration to say that the invention of a process to synthesize nitrogen fertilizer made the modern world possible, as fertilizers boost crop yields as much as 50 percent. Nitrogen fertilizer that isn't taken up by plants boosts input costs to farmers. However, farmers have to make trade-offs between a number of different costs for fuel, equipment, seed, labor, fertilizer, and so forth in order to make a profit, and managing nitrogen fertilizer is usually not at the top of the list for improving the bottom line.

That being said, if it's economic and ecological madness to subsidize the burning of fossil fuels, it's just as barmy to subsidize agriculture in the amount of $300 billion annually. The World Bank reported in 2012 that fertilizer subsidies in India amounted to 2 percent of that country's GDP. Agricultural subsidies clearly encourage farmers to overuse fertilizer, which in turn produces nitrous oxide emissions that harm the ozone layer and raise global temperature.

Real Intergenerational Equity

Comedian Groucho Marx once famously quipped, "Why should I do anything for posterity? What has posterity ever done for me?" Many people are worried about "intergenerational equity" with regard to how global warming will affect future generations. But perhaps Marx had the right question.

Consider that University of Groningen economist Angus Maddison calculates that annual per capita income in real dollars in 1900 in Western Europe was $3,200. Today Western European average incomes average $21,800. Per capita income averaged $4,000 in the United States in 1900. Currently, average American income is $30,500 per capita, according to Maddison's figures. In other words, contemporary Europeans and Americans are around seven times richer than their great-grandparents were three generations ago. The true intergenerational equity question becomes: How much would you have demanded that your much poorer ancestors give up in order to prevent the climate change we are now experiencing? We stand in exactly that same relation to people who will be living in 2100.

Total global GDP in 1900 in real dollars was about $2 trillion. The World Bank calculates that in 2012, global GDP stood at $72 trillion. In other words, global GDP increased by thirty-six times over the past century. Average per capita global real income in 1900 was about $1,300. Dividing the World Bank figure up by the world's population of 7.2 billion, one finds that global average per capita income is around $10,000. Of course, it is not equally distributed among people.

What about the future? If global economic growth continues at around 3 percent per year, total GDP in real dollars would reach $888 trillion in 2100. Many scenarios, including those used by the UN Intergovernmental Panel on Climate Change (IPCC), suggest that world population will stabilize or even fall below 8.5 billion people by 2100. This yields an average income of over $104,000 per person in three generations. So should people living now and making a global average of $10,000 per year be forced to lower their incomes in order to boost the incomes of future generations that in some IPCC scenarios will have incomes in 2100 over $107,000 per capita in developed countries and over $66,000 in developing countries?

Let's take the worst-case scenario devised by British economist Nicholas Stern, in which global warming is so bad that it reduces the incomes of people living in 2100 by 20 percent below what it would otherwise have been without climate change. That implies that global GDP would rise to only $710 trillion by 2100. That would reduce average incomes in 2100 to only $84,000 per person. In other words, people living three generations hence with the worst consequences of climate change would still be more than eight times richer than people living today. Another way to think of that much future climate damage is that it is equivalent to reducing global economic growth from 3 percent to 2.7 percent over the next ninety years.

An Emergency Backup Plan to Cool the Planet

Say that the world has adopted measures that will lead to the massive deployment of low- and no-carbon energy technologies, but it turns out that global warming starts to occur at a much faster pace than climate models had projected. To prepare for just such a situation, some researchers have suggested various geoengineering techniques that could be deployed to slow the rise in temperatures until the low-carbon economy can take hold.

"Prudence demands that we consider what we might do if cuts in carbon dioxide emissions prove too little or too late to avoid unacceptable climate damage," asserted climatologist Ken Caldeira in a 2008 roundtable on geoengineering in *The Bulletin of the Atomic Scientists*. What should we do? "We need a climate engineering research and development plan," declared Caldeira. He added, "We cannot afford a new period of Lysenkoism and allow political correctness to pollute our scientific judgment. Scientific research and engineering development should be divorced from moral posturing and policy prescription." The

National Academy of Sciences released two reports in February 2015 endorsing research into geoengineering strategies that could be deployed to counteract man-made global warming. One is *Climate Intervention: Reflecting Sunlight to Cool the Earth* and the other is *Climate Intervention: Carbon Dioxide Removal and Reliable Sequestration*. The NAS reports argue that both should be pursued.

One proposal involves injecting sulfur particles into the stratosphere, where they would reflect sunlight back into space, thus cooling the planet. There has already been one recent natural experiment that proved this idea could work. In 1991, the Mount Pinatubo volcano in the Philippines cooled the planet when it blasted millions of tons of sulfur particles into the stratosphere; the particulates formed a global haze that lowered average temperatures by about 0.5°C for more than a year.

Researchers and entrepreneurs at Intellectual Ventures have devised a "garden hose to the sky" method for cooling the planet. The firm, founded by polymath and former Microsoft executive Nathan Myhrvold, proposes to hoist an eighteen-mile hose using helium balloons attached every few hundred yards to pump liquefied sulfur dioxide into the stratosphere as a way to mimic the cooling produced by the Pinatubo eruption. The group estimates that setting up five sulfur injection base stations would cost a mere $150 million and cost $100 million per year to operate. If particular areas of the globe—say, the Arctic Ocean and Greenland—are warming up too fast, it might be possible to lower regional temperatures by this means.

Another proposal involves marine cloud whitening, in which hundreds of ships cruise the world's oceans spewing salt water as a mist into the atmosphere. The salt particles would function as cloud condensation nuclei, which would increase the extent and brightness of low level clouds over the oceans. These whitened clouds would reflect sunlight back into space, thus cooling the earth's surface. By one estimate, a fleet of 284 ships spewing salt water into the air at a cost of $1 billion per year would offset about 0.6°C of warming. To reduce future temperatures by 1.9°C, 1,881 vessels would have to be deployed at a cost of $5.8 billion annually.

Some IPCC pathways toward stabilizing the amount of carbon dioxide in the atmosphere explicitly incorporate the development of new technologies that can suck carbon dioxide out of the atmosphere. One such carbon dioxide removal proposal is called bioenergy carbon capture and storage; this involves cultivating plants to absorb carbon dioxide as they grow and then use them as fuel to produce energy. When the plants are burned, the carbon emissions are captured and buried, resulting in negative emissions. Another proposal is direct air capture, a possibility offered by Columbia University researcher Klaus Lackner; this involves using a specific resin that absorbs atmospheric carbon dioxide a thousand times more efficiently than trees do to capture the gas and then store it underground.

Oddly, some environmentalists who profess to be very concerned about the dangers posed by man-made global warming fiercely oppose research on these geoengineering technologies. In 2011, Oxford University researchers planned to conduct what they called the Stratospheric Particle Injection for Climate Engineering (SPICE) experiment in which they merely planned to use balloons to loft a hose to harmlessly spew water droplets into the atmosphere. The experiment was called off under pressure from activists, who denounced it as the pursuit of a "very high-risk technological path" and asserted that "such research is a dangerous distraction from the real need: immediate and deep emissions cuts."

One suspects that the biggest risk that opponents fear is that geoengineering might actually work well. "If humans perceive an easy technological fix to global warming that allows for 'business as usual,' gathering the national (particularly in the United States and China) and international will to change consumption patterns and energy infrastructure will be even more difficult," observed Rutgers University environmental scientist Alan Robock.

Should global temperatures take off steeply, deploying any geoengineering plan would be rife with international political difficulties. For example, the novel climate created by geoengineering would likely shift rainfall patterns, with significant differential impacts on the agricultural sectors of various countries.

Maybe so, but the activists already assert that man-made global warming is a big problem. Since that is so and there's some chance it might come on faster than is currently projected, it is just plain irresponsible to oppose research that could lead to the development an emergency backup cooling system for the planet.

How Much to Insure Against Low Probability Catastrophic Warming?

How much should we pay to prevent the tiny probability of human civilization collapsing? That is the question at the center of an esoteric debate over the application of cost-benefit analysis to man-made climate change. Harvard University economist Martin Weitzman raised the issue by putting forth a Dismal Theorem arguing that some consequences, however unlikely, would be so disastrous that cost-benefit analysis should not apply.

Weitzman contends that the uncertainties surrounding future man-made climate change are so great that there is some nonzero probability that total catastrophe will strike. Weitzman focuses on equilibrium climate sensitivity. Climate sensitivity is defined as the global average surface warming that follows a doubling of atmospheric carbon dioxide concentrations. As has been discussed, the IPCC *Physical Science* report finds that climate sensitivity is likely to be in the range of 1.5° to 4.5°C and *very unlikely* to be greater than 6°C. But very unlikely is not impossible.

Weitzman spins out scenarios in which there could be a 5 percent chance that global average temperature rises by 10°C (17°F) by 2200 and a 1 percent chance that it rises by 20°C (34°F). Considering that the globe's average temperature is now about 15°C (59°F), such massive increases would utterly transform the world and likely wreck civilization. Surely people should just throw out cost-benefit analysis and pay the necessary trillions of dollars to avert this dire possibility, right?

Then again, perhaps Weitzman is premature in declaring the death of cost-benefit analysis. William Nordhaus certainly thinks so, and he has written a persuasive critique of Weitzman's dismal conclusions. First, Nordhaus notes that Weitzman assumes that societies are so risk averse that they would be willing to spend unlimited amounts of money to avert the infinitesimal probability that civilization will be destroyed. Nordhaus then shows that Weitzman's Dismal Theorem implies that the world would be willing to spend $10 trillion to prevent a one-in-100-billion chance of being hit by an asteroid. But people do not spend such vast sums in order to avoid low-probability catastrophic risks. For example, humanity spends perhaps $4 million annually to find and track possibly dangerous asteroids.

Nordhaus also notes that catastrophic climate change is not the only thing we might worry about. Other low-probability civilization-destroying risks include "biotechnology, strangelets, runaway computer systems, nuclear proliferation, rogue weeds and bugs, nanotechnology, emerging tropical diseases, alien invaders, asteroids, enslavement by advanced robots, and so on." As Nordhaus adds, "Like global warming, all of these have deep uncertainty—indeed, they may have greater uncertainty because there are fewer well-understood constants in the biological and technological world than in the geophysical world. So, if we accept the Dismal Theorem, we would probably dissolve in a sea of anxiety at the prospect of the infinity of infinitely bad outcomes." If we applied Weitzman's analysis to our individual lives, none of us would ever get out of bed for fear of dying from a slip in the shower or a car accident on the way to work.

Weitzman's analysis also assumes that humanity will not have the time to learn about any impending catastrophic impacts from global warming. But midcourse corrections are possible with climate change. People would notice if the average temperature began to increase rapidly, for example, and would act to counteract it by cutting emissions, deploying low-carbon technologies, or even engaging in geoengineering. Other low-probability calamities, such as the entire Earth being transformed into strange matter by strangelets produced in high energy physics experiments, don't allow for learning. As Nordhaus dryly notes, "There is no point in revising our views about strangelets in the microsecond after we discover that the calculations of the physicists are wrong." And yet we do not shut down such experiments.

At the end of his critique of Weitzman's Dismal Theorem, Nordhaus investigates what combination of factors would actually produce a real climate catastrophe. He defines a catastrophic outcome as one in which world per capita consumption declines by at least 50 percent below current levels. Since output is projected to grow substantially over the coming century, this implies a decline that is at least 90 percent below the projected baseline. In contrast, the most extreme climate scenario presented by the gloomy *Stern Review* had people living in 2200 making do with only nine times current per capital consumption instead of thirteen times current consumption.

Nordhaus ran a number of scenarios through the Dynamic Integrated Climate-Economy (DICE) model, his integrated assessment model. DICE would produce a catastrophic result only if temperature sensitivity was at 10°C, economic damage occurred rapidly at a tipping point of 3°C, and nobody took any action to prevent the catastrophic chain of events. Interestingly, even when setting all of the physical and damage parameters to extreme values, humanity still had eighty years to cut emissions by 100 percent in order to avoid disaster.

Finally, the question must be asked: Why has no one ever applied a Dismal Theorem analysis to evaluate the nonzero probability that bad government policy will cause a civilization-wrecking catastrophe?

Parsing the Poisonous Politics of Climate Change

The public debate in the United States over climate change science and policy is particularly poisonous. On the one hand, Oklahoma senator James Inhofe denounces climate change science, declaring that "man-made global warming is the greatest hoax ever perpetrated on the American people." On the other, former vice president Al Gore likens those who doubt the seriousness of climate change to odious Holocaust "deniers." What prompts such a level of discord and disrespect? Yale law professor Dan Kahan explains that climate change is not chiefly a fight over science, but is instead one involving a clash of strongly held values. Distressingly, Kahan and his colleagues at the Yale Cultural Cognition Project find in a 2011 study that the more scientifically literate you are, the more certain you are that climate change is either a catastrophe or a hoax.

Many science writers and policy wonks believe that fierce disagreement about issues like climate change is simply the consequence of widespread scientific illiteracy. If this thesis of public irrationality was correct, the authors of the Yale study write, "then skepticism about climate change could be traced to poor public comprehension about science" and the solution would be more science education. In fact, the findings of the Yale researchers suggest more education is unlikely to help build consensus; it may even intensify the debate.

To probe the American public's views on climate change, the Yale researchers conducted a survey of 1,500 Americans in which they asked questions designed to uncover their cultural values, their level of scientific literacy, and their thoughts about the risks of climate change.

The group uses a theory of cultural commitments devised by Aaron Wildavsky, which "holds that individuals can be expected to form perceptions of risk that reflect and reinforce values that they share with others." As noted earlier, the Wildavskyan schema situates Americans' cultural values on two scales, one that ranges from Individualist to Communitarian and another that goes from Hierarchy to Egalitarian. In general, Hierarchical folks prefer a social order where people have clearly defined roles and lines of authority. Egalitarians want to reduce racial, gender, and income inequalities. Individualists expect people to succeed or fail on their own, while Communitarians believe that society is obligated to take care of everyone.

The researchers note that people who hold Individualist/Hierarchical values highly esteem technological innovation, entrepreneurship, and economic growth. Accordingly, they tend to be skeptical of claims about environmental and technological risks and suspect that such claims often amount to little more than unjustifiable excuses for trying to restrict the activities they prize. On the other hand, Egalitarian/Communitarians tend to be morally suspicious of innovation, industry, and commerce, seeing them as the source of unjust disparities in wealth and power. Consequently, they are all too happy to believe claims that those behaviors are risky and impose restrictions on them. In this view, then, Egalitarian/Communitarians would be more worried about climate change risks than Hierarchical/Individualists.

The Yale survey employed a scale in which 1 means no risk and 10 means extreme risk of climate change. The average for the overall sample was a score of 5.7. Hierarchical/Individualists averaged 3.15 points on climate change risk, whereas Egalitarian/Communitarians scored 7.4 on average. The public irrationality thesis predicts that as scientific literacy and numeracy increases, the gap between Hierarchical/Individualists and Egalitarian/Communitarians should lessen. Instead, the Yale researchers found that "among Hierarchical/Individualists science/numeracy is *negatively* [emphasis theirs] correlated with such concern. Hence, cultural polarization actually gets *bigger*, not smaller as science literacy and numeracy increase."

Why does polarization increase with scientific literacy? "As ordinary members of the public learn more about science and develop a greater facility with numerical information, they become more skillful in seeking out and making sense of—or if necessary explaining away—empirical evidence relating to their groups' positions on climate change and other issues," observe the researchers. Confirmation bias, the tendency to search for or interpret information in a way that confirms one's preconceptions, is ubiquitous.

In addition to climate change risks, the Yale researchers surveyed participants for their views on the safety of nuclear power. In this case, greater scientific literacy was associated with reduced concerns about the risks of nuclear power for both groups. However, the gap in perception about the risks of nuclear power between Hierarchical/Individualists and Egalitarian/Communitarians expanded rather than converged as scientific literacy increased. In other words, as scientific literacy increased, Hierarchical/Individualists became much more comfortable with the risks of nuclear power than Egalitarian/Communitarians did. Again, everybody suffers from confirmation bias.

The Yale researchers chalk up this kind of divergence on technological and scientific risks to the pursuit of individual expressive rationality at the expense of collective welfare rationality. Basically, people in both groups are forming beliefs that advance their personal goals and help them get along with the friends and coworkers they interact with on a daily basis. They illustrate the point by observing that "a Hierarchical Individualist in Oklahoma City who proclaims that he thinks that climate change is a serious and real risk might well be shunned by his coworkers at a local oil refinery; the same might be true for an Egalitarian Communitarian English professor in New York City who reveals to colleagues that she thinks that 'scientific consensus' on climate change is a 'hoax.'"

Kahan and his colleagues then argue that what is individually rational when it comes to expressing cultural values becomes collectively irrational in the pursuit of policies aimed at securing society members' health, safety, and prosperity based on what the best scientific evidence reveals about risk and risk abatement. In addition, the researchers note, beliefs about the risks of climate change "come to bear meanings congenial to some cultural outlooks but hostile to others." In this case, Egalitarian/Communitarians, who are generally eager to rein in what they regard as the unjust excesses of technological progress and commerce, see carbon rationing as an effective tool to achieve that goal. This view is distilled in Naomi Klein's book *This Changes Everything: Capitalism vs. the Climate*. Not surprisingly, Hierarchical/Individualists are highly suspicious when proposals involving carbon rationing just happen to fit the cultural values and policy preferences of Egalitarian/Communitarians.

Kahan and his colleagues at the Yale Cultural Cognition Project suggest the Hierarchical/Individualists discount scientific information about climate change because it is strongly associated with the promotion of carbon rationing as the exclusive policy remedy for the problem. They note that other policies that could address climate change might be more acceptable to Hierarchical/Individualists—for example, deploying more nuclear power plants, geoengineering, and developing new technologies to adapt to whatever climate change occurs. While the values of Hierarchical/Individualists steer them toward discounting the dangers of climate change, it is also true that the values of Egalitarian/Communitarians push them to magnify any dangers and to discount the risks that top-down policy interventions pose to the economic well-being of society. Confirmation bias is everywhere.

The Cultural Contradiction of Environmentalist Opposition to Nuclear Power

In one of the more aggravating tales of environmentalist self-preening, former activist and now Vermont Law School professor James Gustave Speth details in his book *Red Sky at Morning: America and the Global Environmental Crisis* how he and others managed to stop the development of no-carbon-emitting fast breeder reactors in the 1970s. For example, as a young attorney for the activist Natural Resources Defense Council (which he cofounded) and the Scientists' Institute for Public Information, Speth filed a key 1973 lawsuit against a government plan to commercialize fast breeders.

Fast breeders are nuclear power plants that can produce more fuel (about 30 percent more) than they use. An additional benefit is that they can produce electricity by burning up highly radioactive nuclear waste and the plutonium removed from nuclear weapons. And it gets better: the radioactive wastes generated by fast breeder reactors after their fuel is recycled decays in only a few hundred years instead of the tens of thousands it takes to render the wastes from conventional reactors harmless. Because the reactors produce more fuel than they use, we would not have to mine any more uranium for thousands of years.

And new fuel-processing technologies have largely allayed concerns that the plutonium produced by fast reactors could be diverted and used to produce nuclear weapons. In other words, fast breeders might have been the ultimate in renewable energy.

The US government projected that as many as two hundred no-carbon-emitting fast breeder reactors would have been generating power by 2000. No one knows for sure if that projection would have come to pass, but had it done so, current US emissions of carbon dioxide would be roughly a third lower than they are now. Thirty years later in his manifesto, Speth asserts that "the biggest threat to our environment is global climate disruption, and the greatest problem in that context is America's energy use and the policies that undergird it." The irony of how his youthful opposition to zero-carbon nuclear energy has contributed to the "context of America's energy use" he now decries evidently escapes Speth.

Nuclear power generation is much safer than coal-power generation. Taking occupational deaths and deaths from pollution into account, one rough estimate finds that coal generation kills about 4,000 times more people than does nuclear generation per unit of power. A study by NASA researchers estimated that by displacing coal generation nuclear power avoided somewhere around 1.8 million deaths from air pollution between 1971 and 2009.

There has been a lot of progress in reactor designs since the 1970s. Westinghouse's new AP1000 reactor is chock-full of all sorts of new safety improvements that can shut down a reactor in crisis with no human intervention. Babcock & Wilcox has designed small modular reactors that could be manufactured and fueled at their plant and then taken by rail to be slotted into already built generating facilities. Once the reactor fuel is spent, it is shipped back to the plant for refueling.

Even more intriguing nuclear technologies are thorium reactors and traveling wave reactors. Thorium is a naturally occurring radioactive element that, unlike certain isotopes of uranium, cannot sustain a nuclear chain reaction. It can, however, be doped with enough uranium or plutonium to sustain such a reaction. Fueled by a molten mixture of thorium and uranium dissolved in fluoride salts of lithium and beryllium at atmospheric pressure, liquid fluoride thorium reactors (LFTRs) cannot melt down. (Strictly speaking, the fuel is already melted.)

Because LFTRs operate at atmospheric pressure, they are less likely than conventional pressurized reactors to spew radioactive elements if an accident occurs. In addition, an increase in operating temperature slows down the nuclear chain reaction, stabilizing the reactor. And LFTRs are designed with a salt plug at the bottom that melts if reactor temperatures somehow do rise too high, draining reactor fluid into a containment vessel, where it essentially freezes.

A 2009 NASA report notes that the radioactivity in wastes from LFTRs "would decay to background levels in less than 300 years, as contrasted to over 10,000 years for currently used reactors, thus obviating the need for long term storage, such as at Yucca Mountain." In fact, LFTRs can burn the long-lived plutonium and other nuclear wastes produced by conventional reactors as fuel, transmuting them into much less radioactive and harmful elements. No commercial thorium reactors currently exist, but China is working on a project that aims to develop them within ten years.

The US company TerraPower's traveling wave reactors are designed to run on what is now essentially nuclear waste. The unique fuel cycle of traveling wave reactors use U-238, often referred to as depleted uranium. The United States has more than 700,000 metric tons of depleted uranium in storage. The amazing fact is that burning those stores of depleted uranium in traveling wave reactors could supply the United States with electricity for thousands of years. TerraPower estimates that burning global stores of depleted uranium (about 1.5 million metric tons) could supply 80 percent of the world's population with the amount of electricity Americans use per capita today for the next thousand years.

How do these reactors work? A traveling wave needs to be ignited only once, using just a bit of U-235 or plutonium to jump-start a chain-reaction wave of neutrons that continuously converts U-238 into plutonium-239. The traveling wave reactor also reduces nuclear weapons proliferation risks, since its fuel cycle would eliminate the need for numerous uranium processing plants. As Charles W. Forsberg, executive director of the Nuclear Fuel Cycle Project at the Massachusetts Institute of Technology, has quipped, the traveling wave fuel cycle "requires only one uranium enrichment plant per planet."

The plutonium burns itself up as it sustains a further chain reaction by transforming depleted uranium into more plutonium. In other words, a traveling wave reactor produces plutonium and uses it up at once, which means that, unlike fuel in conventional reactors, there is very little left over that could be diverted for weapons production. The traveling wave moves through the reactor core at a rate of about a centimeter per year, somewhat like a cigarette burns from tip to filter. Or think of it as two waves, a breeding wave that produces plutonium which is followed close behind by a burning wave that consumes the plutonium. The core is cooled with liquid sodium and the heat is drawn off to produce steam to drive electric generators. TerraPower expects that traveling wave reactors will be sealed and will operate for fifty to a hundred years without refueling or removing any fuel from the reactor.

One of the bitter jokes popular among frustrated aficionados of nuclear fusion power is that practical fusion energy is only thirty years away and always will be. In October 2014 researchers at aerospace giant Lockheed Martin confidently announced that they had made a technological breakthrough that would enable them to build and test a prototype compact fusion reactor in a year and begin deploying them in ten years. Essentially, the Lockheed researchers claimed to have figured out how to confine the hot plasma needed for fusion in a much smaller and less finicky magnetic bottle than the conventional tokamak reactor. How much smaller? Ten times smaller, one that would fit in the back of a large truck. Such a 100-megawatt compact fusion reactor could supply enough electricity to run a small city of 100,000 people. It might even be used to power aircraft. The reactor could run for a year by fusing fifty-five pounds of deuterium and tritium to produce the heat needed to drive generators. Fusion reactors produce much less radioactive waste than conventional fission reactors and no greenhouse gases to warm the planet's atmosphere.

Why Not Deploy Current Renewable Power Technologies Now?

"We have the tools—the technologies, the resources, the economic models—to deliver cost-effective climate solutions at scale," testified K. C. Golden of the US-based NGO Climate Solutions before the Senate Committee on Environment and Public Works in July 2013. Friends of the Earth issued a similar statement in September 2013: "We have the technology we need [to address climate change] and we know what needs to happen. We just need to get politicians to do it." Tove Maria Ryding, coordinator for climate policy at Greenpeace International, sounded the same note in 2012: "We have all the technology we need to solve the [climate] problem while creating new green jobs."

The implication is that humanity could deploy a suite of currently available zero-carbon energy production technologies and energy efficiency improvements to avert the impending climate catastrophe. And the idea has been around for a while. Back in 2008, Al Gore urged America "to commit to producing 100 percent of our electricity from renewable energy and truly clean carbon-free sources within 10 years," a goal that he pronounced "achievable, affordable and transformative." His plan was possible, he explained, because the price of the technologies needed to produce no-carbon electricity—solar, wind, and geothermal—were falling dramatically.

As it happens, America did not take up the former vice president's challenge. In 2014, solar, geothermal, and wind energy generated 0.46, 0.39, and 4.28 percent, respectively, of electric power in the United States.

Was Gore right seven years ago? And are the folks at Greenpeace, Friends of the Earth, and Climate Solutions right now that the no-carbon energy technologies needed to replace fossil fuels are readily available and ready to go?

Not really, concludes a November 2013 report, "Challenging the Clean Energy Deployment Consensus," by the Washington, DC-based Information Technology and Innovation Foundation (ITIF). Such plans, the study argues, "are akin to attempting large-scale moon colonization using Apollo-age spacecraft technology." Such a feat may be technically feasible, but only at vast expense.

Think of the issue this way: Would you rather drive a 1913 Model T Ford or a 2013 Ford Fiesta? They cost about the same amount of money in inflation-adjusted dollars. One way to interpret the ITIF report is that the advocates of immediately deploying current zero-carbon energy production technologies are essentially arguing that we should now all be driving Model T Fords.

To get some idea of what would be involved in "repowering" America using only the currently available zero-carbon technologies, let's delve into one of the more ambitious of the studies that the ITIF folks criticize. In a 2011 paper, the Stanford engineer Mark Jacobson and the University of California at Davis transportation researcher Mark Delucchi calculated what it would take to produce all the energy (not just electric power generation) to fuel the entire world using zero-carbon sources by 2030. They also calculate what renewable sources of energy would be needed to power just the United States. They conclude that this would require 590,000 5-megawatt wind turbines, 110,000 wave devices, 830 geothermal plants, 140 new hydroelectric dams, 7,600 tidal turbines, 265 million rooftop solar photovoltaic systems, 6,200 300-megawatt solar photovoltaic power plants, and 7,600 300-megawatt concentrated solar power plants.

Let's adjust those figures to take into account the fact that we currently use 40 percent of primary energy to generate electricity. Making the heroic assumption that Americans will consume no more electricity in 2030 than they do today, what would it take to "repower" the country's 1,000-gigawatt electric generation sector entirely in zero-carbon renewable energy sources? Keep in mind that the total asset value of the entire US electrical system, including generation, distribution, and transmission, amounted to $800 billion in 2003.

Well, first we would have to install 15,000 new wind turbines, 155 solar photovoltaic, and 190 concentrated solar power plants each year. In 2012, the US wind industry installed a record 13 gigawatts of rated generating capacity; construction of 15,000 5-megawatt turbines annually for the next sixteen years entails a fivefold jump in the installation rate. Building 13 gigawatts cost $25 billion, which implies an increase to $125 billion annually, reaching a total cost over the next sixteen years of $2 trillion. And that's just for wind power.

In 2012, the world's largest solar photovoltaic plant came online in Arizona at Agua Caliente. That facility, rated at 250 megawatts of generation capacity, cost $1.8 billion to build. Achieving the zero-carbon repowering goal implies constructing 155 of these each year for the next sixteen years. The costs would amount to roughly $280 billion annually, for a total of $4.5 trillion. The United States is also home to the world's largest concentrated solar power plant at Ivanpah, California. That 372-megawatt plant cost $2.2 billion to build, which implies spending about $440 billion annually for 190 such plants, adding up over sixteen years to roughly $7 trillion.

That's just to build enough *rated* zero-carbon generation capacity to replace what we have now. As the ITIF study makes clear, most renewable power sources are highly variable in their production. The deploy-now crowd hopes that somebody will invent some way to store electricity so that it could make up for shortfalls when the sun doesn't shine or the wind fails to blow.

A 2013 study analyzed by the ITIF researchers solves this renewable energy storage problem by oversizing—that is, by building two to three times more generating capacity than would be necessary if they could operate near their rated capacity all of the time. This suggests that at the low end of this estimate would raise the estimated costs in the repowering scenario by 2030 to $4 trillion for wind generation and to more than $23 trillion the total solar portion.

Admittedly my preliminary rough calculations assume that costs for constructing zero-carbon energy sources do not fall over the next sixteen years. The price of the Model T Ford, introduced in 1908 at the price of $850 ($22,000 in 2013 dollars), fell from $550 ($13,000 in 2013 dollars) to $260 ($3,500 in 2013 dollars) by its last year of production in 1927. Assuming that the costs of installing current versions of zero-carbon energy production technologies fell as much immediately, the total costs for would still amount to roughly $7 trillion by 2030.

The ITIF analysis alternatively adds up all of the costs in the Jacobson/Delucchi paper to estimate that weaning Americans off fossil fuels entirely by 2030 would add up to a total of $13 trillion—that is to say, 5 percent of each year's GDP over the next sixteen years. The upshot is that this repowering would cost each American household an additional $5,664 per year until 2030.

Are Americans really willing to shell out that much cash for zero-carbon energy? The ITIF report observes that a 2011 poll found that Americans were willing to pay just under $10 per month ($120 per year) more for electricity generated by renewable sources. In addition, half of Americans can choose to pay about 10 percent more to purchase electricity generated from renewable sources, but only 1 percent actually do so.

These calculations are just for the United States. Somewhere around 1.3 billion people around the world still do not have access to electricity. Taking the Jacobson and Delucchi figures for the world, the total cost to completely eliminate fossil fuels by 2030 would amount to $100 trillion—that is to say, 8 percent of global annual GDP. The global cost per household per year would amount to $3,571. The nearly 3 billion people who live on less than $2,000 per year simply cannot pay the prices needed to deploy current versions of renewable power technologies.

Despite the foregoing analysis, technological innovation and competitive markets may yet come to the rescue during the coming decades.

Unlimited Free Solar Power?

"Despite the skepticism of experts and criticism by naysayers, there is little doubt that we are heading into an era of unlimited and almost free clean energy," the Stanford technology maven Vivek Wadhwa declared in *The Washington Post* in September 2014. The technology that most inspires his enthusiasm is solar energy—and while solar isn't close to "almost free" yet, it is indeed getting cheaper. The prices of solar photovoltaic (PV) modules have fallen steeply by more than 80 percent since 2008.

This trajectory seems to be following Swanson's Law, named for Richard Swanson, the founder of US solar-cell manufacturer SunPower. Swanson suggested that the cost of the photovoltaic cells falls by 20 percent with each doubling of global manufacturing capacity. The pattern is a product of constantly improving manufacturing processes: more automation, better quality control, materials reduction, and so forth.

But how plausible is Wadhwa's prediction that solar power will be unlimited and nearly free? To get a handle on solar's future, let's look at a measure called the *levelized cost of energy*. This takes into account the capital costs, fuel costs, operations and maintenance costs, debt and equity costs, and plant utilization rates for each type of electric power generation. Many different groups have tried to calculate and compare the levelized costs for building, operating, and financing coal, natural gas, nuclear, hydro, solar, wind, geothermal, and biomass plants.

Let's start with the levelized cost analysis that is the most bullish with respect to solar photovoltaic. In September 2014, the financial advisory firm Lazard reckoned that the levelized unsubsidized cost of utility-scale solar PV is as low as $72 per megawatt-hour. (A megawatt-hour is roughly equivalent to the amount of electricity used by 330 houses during one hour.) Lazard projects that these costs will drop to $60 per megawatt-hour by 2017. Meanwhile, the low end of natural gas generation is now $61 per megawatt-hour; for coal generation, it's $66 per megawatt-hour; and for nuclear, it's $124 per megawatt-hour. With the current US tax breaks, the low-end solar PV utility-scale costs is $56 per megawatt-hour. Even so, George Bilicic, a vice chairman of Lazard, concluded that utilities "still require conventional technologies to meet the energy needs of a developed economy, but they are using alternative technologies to create diversified portfolios of power generation resources."

Every couple of years the Electric Power Research Institute, a nonprofit think tank sponsored by the electric power generation industry, issues a report on the levelized cost of energy for various power generation technologies. Its *Integrated Generation Technology Options 2012* report calculates the low-end levelized cost for solar PV next year at $107 per megawatt-hour. For natural gas, coal, and nuclear, the low-end costs are $33, $62, and $85 per megawatt-hour, respectively.

The institute calculates that by 2025, the low-end levelized costs of solar PV will fall to $81 per megawatt-hour. By that time, the institute expects that coal plants will be required to capture their carbon emissions, so the levelized cost of coal will be $102 per megawatt-hour. Natural gas plants without carbon capture will face levelized costs of $44 per megawatt-hour. The report cautions that its calculations with respect to renewable energy generation do not take into account additional costs, such as backup generation or integration into the electric power grid. If included, such costs would substantially raise the levelized costs of renewable energy generation technologies.

One other authoritative analysis is the *Annual Energy Outlook* published by the US Energy Information Administration (EIA). In its 2014 report, the agency reckons that in 2019, the low-end cost of solar PV will be $101 per megawatt-hour. Conventional coal, nuclear, and natural gas levelized costs stand correspondingly at $87, $92.60, and $61.10 per megawatt-hour.

To judge from these estimates, the era of unlimited, nearly free solar power has certainly not yet arrived. But things are moving quickly. As recently as 2011, the EIA did not even bother trying to calculate levelized solar PV costs. In that year's report, the agency projected that the country would have an installed solar PV capacity of 8.9 gigawatts by 2035. As of the second quarter of 2014, the figure was already 15.9 gigawatts.

In 2008, global production capacity of solar cells/modules amounted to 7 gigawatts. Capacity is now projected to be 85 gigawatts in 2016. This rate of increase suggests a manufacturing capacity doubling time of about two years. As capacity ramped up, Lazard reports that the levelized costs fell from $323 per megawatt-hour in 2009 to $72 now. If Swanson's Law proves true, the levelized cost solar PV could be expected to fall to around $24 per megawatt-hour in the next ten years. That would not be too cheap to meter, but it would cost far less than any of the current forecasts for fossil fuel electric power generation technologies.

Of course, this rough projection does not take into account the huge issue of intermittency (the sun doesn't always shine) that makes solar power problematic as a baseload source of electricity. However, potentially disruptive innovations like the solar subcell developed by German Fraunhofer Institute for Solar Energy Systems that can turn 44.7 percent of sunlight that strikes it into electricity or Sakti3's new high-capacity battery that the Michigan-based company claims offers double the energy density of current lithium-ion technology at a fifth the cost could accelerate the wider adoption of solar power.

Will Wadhwa's prophecy come true? Perhaps not, but wagering against human ingenuity has always been a bad bet.

Let us turn now to how to consider human ingenuity might be better harnessed to solve the climate/energy conundrum than trying to impose various forms of carbon rationing on the world.

The Emerging Climate and Energy Consensus

"The Kyoto Protocol is dead. There will be no further global treaties that set binding limits on the emissions of greenhouse gases (GHG) after Kyoto runs out in 2012." That's what I wrote back in 2004 when I was reporting on the 10th Conference of the Parties to the UNFCCC in Buenos Aires, Argentina. I also cited the prediction of Taishi Sugiyama, a senior researcher at Japan's Central Research Institute of Electric Power Industry, who flatly stated that Kyoto signatories Canada, Japan, and Russia would withdraw from the treaty after 2012. As noted earlier, Sugiyama's prediction has come true.

Instead of UN carbon-rationing schemes, Sugiyama recommended in 2004 that a clean energy technology-push approach be formalized in a Zero Emissions Technology Treaty. Such a treaty would have greater appeal because it avoids the inevitable conflicts over allocating emissions targets and because most countries recognize the importance of long-term technological progress. Sugiyama presciently argued that a global cap-and-trade system is way too premature for developing countries to join because effective low-cost ways to cut carbon emissions simply don't exist. "I cannot imagine a cap-and-trade system over the whole globe without low-cost energy and emissions control technologies," said Sugiyama. Ten years later, Sugiyama's insight that the Kyoto Protocol is a dead end and that the best way to address man-made global warming is to develop clean energy sources so that they become cheaper than fossil fuels is emerging as the new energy technology consensus.

In May 2014, former undersecretary for global affairs Timothy Wirth, who was the Clinton administration's lead negotiator for the Kyoto Protocol, and former South Dakota senator Thomas Daschle conceded that "the international community should stop chasing the chimera of a binding treaty to limit CO_2 emissions." They note that more than two decades of UN climate negotiations have failed because "nations could not agree on who is to blame, on how to allocate emissions, or on projections for the future." Both firmly believe that man-made global warming poses significant risks to humanity.

To address those risks, Wirth and Daschle now advocate that the climate negotiators adopt a system of "pledge and review" at the 2015 Paris conference of the parties to the UNFCCC. In such a scheme, nations would make specific pledges to cut their carbon emissions, to adopt clean energy technologies, and to wring more GDP out of each ton of carbon emitted. The parties would review their progress toward reducing greenhouse gas emissions every three years and make further pledges as necessary to achieve the goal of keeping the increase in average global temperature under 2°C. Since there would be no legally binding targets, there would be no treaty that would require politically difficult ratification. If insufficient progress is being made by 2020, they argue that countries should consider adopting a globally coordinated price on carbon.

Wirth and Daschle note that the markets for renewable energy, especially wind and solar, as well as natural gas, the least carbon-intensive fossil fuel, are expanding. Crucially the two believe that the 2015 UN climate change conference in Paris should aim to "help accelerate the pace of technological adoption and change, toward the day when the cleanest energy sources are also the cheapest and thus become dominant." Clearly they have joined the emerging consensus that schemes to prevent climate change by rationing carbon—for example, by imposing a cap-and-trade scheme or taxation—are doomed to failure.

Why failure? Because of the "iron law of climate policy," argues University of Colorado political scientist Roger Pielke Jr. Pielke's iron law declares that "when policies focused on economic growth confront policies focused on emissions reductions, it is economic growth that will win out every time." People and their governments are very reluctant to give up the immediate benefits of economic growth—more goods and services, jobs, better education, and improved health—that access to modern fuels make possible in order to avert the distant harms of climate change.

Make Clean Energy Cheaper Than Fossil Fuels

"The paramount goal of climate policy should be to make the unsubsidized cost of clean energy cheaper than fossil fuels so that all countries deploy clean energy because it makes economic sense," is how the Information Technology and Innovation Foundation sums up the new consensus. This perspective is also endorsed by many other policy groups, including the Breakthrough Institute and the Brookings Institution.

"Societies that are able to meet their energy needs become wealthier, more resilient, and better able to navigate social and environmental hazards like climate change," correctly notes the Breakthrough Institute's 2014 *Our High Energy Planet* report. Keeping people in developing countries in comparative energy poverty will only slow energy innovation. "The way we produce and use energy will become increasingly clean not by limiting its consumption, but by using expanded access to energy to unleash human ingenuity in support of innovating toward an equitable, low-carbon global energy system," asserts the Breakthrough Institute report.

The first plank of the new consensus is that it is wrong to try to restrain the growth of greenhouse gas emissions by denying adequate access to modern fuels to the poor. For example, the Breakthrough Institute report rejects the International Energy Agency's anemic recommendation that annual access to 100 kilowatt-hours of electricity per person is sufficient. That is the amount of electricity that the average American burns in three days and the average European consumes in five days. One reasonable threshold might be 8,000 kilowatt-hours, which is the quantity that the average Japanese citizen uses in a year.

Second, activist opposition to safe hydraulic fracturing to release vast quantities of natural gas trapped in deep underground shale formations is counterproductive. Burning natural gas releases about half the carbon dioxide that burning coal does. In fact, the 2013 IPCC *Physical Science* report identifies power generation using natural gas as a "bridge technology" that can be deployed now. Consequently, the IPCC report notes, "Greenhouse gas emissions from energy supply can be reduced significantly by replacing current world average coal-fired power plants with modern, highly efficient natural gas combined-cycle power plants." Coal-fired electric power plants that emit lots of carbon dioxide are largely being shut down in the United States because they cost more than plants that emit far less carbon dioxide by burning cheap natural gas produced through fracking.

Third, environmentalist hostility to all forms of nuclear power is similarly perverse. Generating electricity using nuclear power emits almost no greenhouse gases while assuring a stable supply of baseload power. Activist resistance to nuclear power may be lessening. No one would accuse climate researchers James Hansen, Kerry Emanuel, Ken Caldeira, and Tom Wigley of excessive moderation when it comes to banging the climate crisis drum. In November 2013 the four joined the new consensus by issuing an open letter challenging the broad environmental movement to stop fighting nuclear power and embrace it as a crucial technology for averting the possibility of a climate catastrophe by supplying zero-carbon energy. The letter point-blank states that "continued opposition to nuclear power threatens humanity's ability to avoid dangerous climate change." The four add, "While it may be theoretically possible to stabilize the climate without nuclear power, in the real world there is no credible path to climate stabilization that does not include a substantial role for nuclear power."

The fourth and most provocative plank of the new energy technology consensus is that government research and development spending on zero-carbon forms of energy supply must be dramatically ramped up. "Robust government support, including significant investment for clean energy research, development, and demonstration (RD&D), is necessary to make energy technologies cheaper than fossil fuels," argues the ITIF.

Those of us who appreciate the power of competition and market incentives to call forth new technologies and drive down prices must recognize that governments have been massively meddling in energy markets for more than a century. Consequently, it's really impossible to know what the actual price of energy supplies would be in a free market. Notorious examples include the attempts at petroleum cartelization by the Organization of Petroleum Exporting Countries (OPEC) and Russia's constant jiggering of the price of natural gas sold to European countries. In many countries, electricity is generated and distributed by government agencies that are not accountable to consumers.

For example, in the United States, a system of electricity regulation that chiefly benefited producers at the expense of consumers was established at end of the nineteenth and the beginning of the twentieth centuries. And no segment of energy supply has gone unmolested by the federal government. A comprehensive 2011 analysis of US federal government energy tax, regulatory, and research and development incentives finds that they have amounted to more than $837 billion (2010 constant dollars) since 1950. Of that amount, $153 billion was spent on energy research and development. By 2010, nuclear energy had received $74 billion in R&D funding, coal, $36 billion, and renewables, $24 billion. Federal R&D for oil, natural gas, and geothermal totaled

$21 billion. These subsidies undoubtedly distorted both the supply and demand for these forms of energy, thus masking the actual comparative costs and benefits of each.

The better course would be to establish a level playing field by eliminating all energy subsidies and incentives and letting the cheapest technologies developed by innovators win in the marketplace. Proponents of markets must continue to push policy in this direction, but given the history of pervasive government intervention in energy markets, it is unlikely that governments will suddenly step back and allow markets to decide how to innovate and produce energy in the future. Energy production, especially for electricity, approximates a government-sanctioned monopoly that has the unfortunate side effect of stifling private innovation in energy production technology. Given that situation, the new consensus in favor of government-subsidized energy production research and development that aims to make zero-carbon energy supplies cheaper than fossil fuels looks like the least bad likely policy option for addressing concerns about climate change.

How much do proponents of the new consensus want to spend on clean energy R&D? The ITIF report suggests investing $70 billion per year globally, which amounts to less than 13 percent of the funds spent worldwide on fossil fuel subsidies now. In addition, $70 billion in R&D funding represents only about a quarter of the $254 billion spent in 2013 on deploying currently expensive and technologically clunky renewable power technologies. Given those figures, the ITIF's estimate of what it would take to develop cheap zero-carbon technologies looks like a bargain.

The Climate Change Bottom Line

Despite the current pause in global warming and the real failings in climate computer model projections, the balance of the scientific evidence suggests that man-made climate change could become a significant problem by the end of this century. As we have seen, political progressives and environmentalists like Naomi Klein fervently promote the "climate crisis" as a pretext for radically transforming the world's economy in ways that ratify their own ideological predilections. Thus they advocate the imposition of vast top-down regulatory schemes that ultimately amount to various forms of carbon and energy rationing.

As a response, lots of supporters of free markets and economic growth tend to underplay the science that suggests the possibility that continued unrestrained emissions of greenhouse gases could have really undesirable effects on the planet's climate by the end of the century. Why? Because they have fallen for the false dilemma posed by progressive environmentalists of supposedly having to choose between economic growth and averting the possibility of disruptive climate change. A far better strategy for challenging radical progressive proposals is to advocate policies that further enable market-driven advances in science and technology to cut through the climate/energy conundrum. Among other things, this would include eliminating all energy subsidies, most especially those to fossil fuels.

It is surely the case that if one wants to help future generations deal with climate change, the best policies are those that encourage rapid economic growth. This would endow future generations with the wealth and superior technologies necessary to handle whatever comes at them, including climate change. In other words, in order to truly address the problem of climate change, responsible policymakers should select courses of action that move humanity from a slow-growth trajectory to a high-growth trajectory, especially for the poorest developing countries. Whatever slows down economic growth will also slow down environmental cleanup and renewal.

7

Is the Ark Sinking?

WHEN I WAS A BOY, MY FASCINATION WITH the plight of the whooping cranes was kindled by the book *The Whooping Crane* by National Audubon Society ornithologist Robert P. Allen. Allen was the man who is most responsible for bringing America's tallest bird back from the brink of extinction. The total population in the wild had fallen from an estimated 10,000 before European settlement to just 15 birds by the 1930s. I was so taken by Allen's intrepid and passionate story that during a mid-1960s visit to my Texas grandparents I whined and wheedled my parents into taking me to the San Antonio Zoo to see the two captive whoopers, Rosie and her mate, Crip. Fortunately, I was not viewing the last representatives of a species on its way out, but one on its way back. The good news is that the wild migratory population has recovered to around 280 birds and some 290 others are captive or part of reintroduction efforts. Biologists believe that nurturing the species to

1,000 birds with 250 breeding pairs would pull the whooping cranes safely back from the threshold of extinction.

While the fortunes of the whoopers may be improving, many biologists and conservationists are urgently warning that humanity is on the verge of wiping out hundreds of thousands of other species in this century. "A large fraction of both terrestrial and freshwater species faces increased extinction risk under projected climate change during and beyond the 21st century," states the 2014 IPCC *Adaptation* report. "Current rates of extinction are about 1000 times the likely background rate of extinction," starkly asserts a May 2014 review article in *Science* by Duke University biologist Stuart Pimm and his colleagues. "Scientists estimate we're now losing species at 1,000 to 10,000 times the background rate, with literally dozens going extinct every day," warns the Center for Biological Diversity. The CBD adds, "It could be a scary future indeed, with as many as 30 to 50 percent of all species possibly heading toward extinction by mid-century." Eminent Harvard University biologist E. O. Wilson agrees. "We're destroying the rest of life in one century. We'll be down to half the species of plants and animals by the end of the century if we keep at this rate." University of California at Berkeley biologist Anthony Barnosky similarly notes, "It looks like modern extinction rates resemble mass extinction rates." Assuming that species loss continues unabated, Barnosky adds, "The sixth mass extinction could arrive within as little as three to 22 centuries."

The Sixth Mass Extinction?

Barnosky is comparing contemporary estimates of species loss to the five prior mass extinctions that occurred during the past 540 million years in which around 75 percent of all then-living species died off each time. The most famous extinction episode—likely triggered by an asteroid crashing into the earth—killed off the dinosaurs 65 million years ago. The asserted cause of the sixth extinction event is human activity, chiefly the result of cutting down forests and warming the planet.

About 1.9 million species have so far been described by researchers, whose estimates of the total number of species on the planet range from 3 to 10 million.

Let's assume 5 million species. If Wilson is right that half could be gone by the middle of this century, that implies that species are disappearing at a rate of 71,000 per year, or just under 200 per day. Contrast this implied extinction rate with Pimm and his colleagues, who estimate that the background rate of extinction without human influence is about 0.1 species per million species years. This means that if one followed the fates of 1 million species, one would expect to observe about one species going extinct every 10 years. Their new estimate is 100 species going extinct per million species years. So if the world contains 5,000,000 species, then that suggests that 500 are going extinct every year. Obviously, there is a huge gap between Wilson's off-the-cuff estimate and Pimm's more cautious calculations, but both assessments are troubling.

Earlier Extinction Predictions

However, this is not the first time that biologists have sounded the alarm over purportedly accelerated species extinctions. In 1970, Dr. S. Dillon Ripley, secretary of the Smithsonian Institution, predicted that in twenty-five years, somewhere between 75 and 80 percent of all the species of living animals would be extinct. That is, 75 and 80 percent of all species of animals would be extinct by 1995. Happily, that did not happen. In 1975, Paul Ehrlich and his biologist wife, Anne Ehrlich, predicted that "since more than nine-tenths of the original tropical rainforests will be removed in most areas within the next thirty years or so, it is expected that half of the organisms in these areas will vanish with it." It's now nearly forty years later and nowhere near 90 percent of the rain forests have been cut and no one thinks that half of the species inhabiting tropical forests have vanished.

In 1979, Oxford University biologist Norman Myers stated in his book *The Sinking Ark* that 40,000 species per year were going extinct and that 1 million species would be gone by the year 2000. Myers suggested that the world could "lose one-quarter of all species by the year 2000." At a 1979 symposium at Brigham Young University, Thomas Lovejoy, who is the former president of the H. John Heinz III Center for Science, Economics, and the Environment, announced that he had made "an estimate of extinctions that will take place between now and the end of the century. Attempting to be conservative wherever possible, I still came up with a reduction of global diversity between one-seventh and one-fifth." Lovejoy drew up the first projections of global extinction rates for the *Global 2000 Report to the President* in 1980. If Lovejoy had been right, between 15 and 20 percent of all species alive in 1980 would be extinct right now. No one believes that extinctions of this magnitude have occurred over the last three decades.

What did happen? As of 2013, the International Union for the Conservation of Nature (IUCN) lists 709 known species as having gone extinct since 1500. A study published in *Science* in July 2014 reported that among terrestrial vertebrates, 322 species have become extinct since 1500. That being noted, the IUCN Red List records 6,451 species as endangered and 4,286 as critically endangered. Species are considered to be endangered if, among other findings, they number fewer than 2,500 mature individuals, their habitat encompasses fewer than 5,000 square kilometers, and/or their population size has been reduced by more than 70 percent over the last ten years or three generations, whichever is the longer. They are deemed critically endangered if they number fewer than 250 mature individuals, their range is less than 100 square kilometers, and/or their population has been reduced by more than 90 percent over the last ten years or three generations, whichever is longer.

In September 2014, the World Wildlife Fund published its *Living Planet Index 2014* report, which alarmingly calculates that the Earth is home to about half the number of vertebrates (mammals, birds, reptiles, amphibians, fish) that it hosted in 1970. Let's be clear: The report is *not* saying that half of vertebrate species have gone extinct, but that the overall number of wild vertebrates have declined by half. The trend is calculated using a complicated system for weighting the declines in various vertebrate species populations. The report also finds that 37 percent of the population declines result from direct exploitation (for example, overfishing and hunting); 31.4 and 13.4 percent are from habitat degradation and destruction (for example, cutting down tropical forests).

In an effort to deal with the threat of species extinction, in 1973 the United States adopted the Endangered Species Act, with the goal being to prevent the extermination of native species. The United States is home to approximately 200,000 species. In 2014, there were 1,529 domestic and 625 foreign species listed as either endangered or threatened under the Endangered Species Act.

A 2004 report by the Center for Biological Diversity lists 108 species as having gone extinct since the adoption of the Endangered Species Act. The researchers found that while 23 species became extinct after they were placed on the endangered species list, 85 species that died off never made it onto the list. The CBD list of US extinctions underlines the reality that most extinctions occur on oceanic islands—for example, Hawaii, Guam, and Puerto Rico—and in freshwater streams. The relatively small size and isolation of islands and freshwater streams render their endemic species especially vulnerable to being wiped out. In addition, lots of island species have lost their wariness of predators and consequently are devastated when mainland carnivores and omnivores such as rats and pigs are introduced. As a result, most of the species listed as going extinct in the CBD report were island endemics or denizens of freshwater streams (mostly mollusks). Interestingly, the 2004 CBD report lists the giant Palouse earthworm as having been extinct since 1978. The good news is that it was rediscovered in 2008 and was found to be so abundant that the US Fish and Wildlife Service declined in 2011 to list it as endangered.

Still, humanity is quite capable of wiping out species. Just as the last ice age was ending, our hunter-gatherer ancestors spread across the world, killing off megafaunal populations already stressed by climate change. Some 178 mammal species weighing more than a hundred pounds disappeared, drastically reducing the total mammalian biomass of the planet. For example, after humans arrived in North America, more than thirty different groups of large mammals, including horses, camels, mammoths, and mastodons, disappeared. In South America, 100 percent of mammals weighing more than a ton, including ground sloths, armadillo-like glyptodonts and rhinoceros-like toxodons, and 80 percent of those weighing more than a hundred pounds went extinct. Total mammalian biomass did not recover until just before the Industrial Revolution, when the post-ice-age losses were finally offset by the collective weight of the populations of humans and our domesticated animals. A 2013 study estimates that Polynesian wayfarers caused the extinction of 1,300 species of birds as they colonized the isolated islands of the Pacific Ocean. The arrival of Europeans killed off an additional 40 Pacific island bird species.

As we've seen from the IUCN list, biologists are not actually counting the number of species that are going extinct. As the example of the giant Palouse earthworm shows, it is really difficult to be sure when the last individuals of a species die off. So how do biologists come up with their shocking estimates of the number of species that they believe are likely to go extinct before the end of this century?

Calculating Extinctions

For the most part, the dire extinction estimates cited earlier are based on computer model calculations using the species-area-curve relationship derived from the theory of island biogeography. In the 1960s, Harvard University biologists E. O. Wilson and Robert MacArthur devised that theory, which basically predicts that the bigger the island, the more species it can support. This relationship is captured in the species-area curve. As Wilson explained it, in general if an ecosystem is reduced by 90 percent, the number of different species it can sustain is cut by 50 percent. For example, the Ehrlichs simplistically extrapolated from this crude species-area-curve relationship to make their wrong prediction back in 1975 that half of all tropical species would be extinct by 2000.

More recent research is questioning the calculations made on the basis of the species-area-curve relationship. For example, in a 2011 article in *Nature*, "Species-Area Relationships Always Overestimate Extinction Rates from Habitat Loss," researchers concluded that "extinctions caused by habitat loss require greater loss of habitat than previously thought" and that reliance on the species-area curve overestimates species loss by as much as 160 percent.

The IPCC's 2014 *Adaptation* report notes: "Models project that the risk of species extinctions will increase in the future due to climate change, but there is low agreement concerning the fraction of species at increased risk, the regional and taxonomic focus for such extinctions and the timeframe over which extinctions could occur." That is to say, the computer models that researchers use to try to estimate the effects of climate change on species don't agree on how many species are at risk, where they are at risk, which species are at risk, and how long it would take before species went extinct. As a result, the report finds that "model-based estimates of the fraction of species at substantially increased risk of extinction due to 21st century climate change range from below 1% to above 50% of species in the groups that have been studied." Another interesting observation in that report is that "evidence from the paleontological record indicating very low extinction rates over the last several hundred thousand years of substantial natural fluctuations in climate—with a few notable exceptions such as large land animal extinctions during the Holocene—has led to concern that forecasts of very high extinction rates due entirely to climate change may be overestimated." Furthermore, the *Adaptation* report notes, "The limited number of studies that have directly compared land use and climate change drivers have concluded that projected land use change will continue to be a more important driver of extinction risk throughout the 21st century." In other words, the biggest peril faced by species is not climate change but how human beings use and alter landscapes.

Since how people use land and water is the critical factor in protecting species from extinction, looking to the future there is good news with regard to strongly positive trends in population, farmland, urbanization, protected areas, and wealth. As we've seen in earlier chapters, human population will most likely peak in this century and begin to fall. Second, average wealth will also increase substantially, which will generate more demand for environmental quality, including the expansion and protection of wild areas.

Expanding Protected Land and Seascapes

In fact, the expansion of protected areas is already happening at a remarkably fast pace. The World Bank notes that protected areas have nearly doubled from 8.5 percent in 1990 to 14.3 percent in 2012 of the world's total land area. That's an area twice the size of the entire United States. Marine protected areas have increased from 4.7 percent of territorial waters in 1990 to 10 percent in 2012. Under the Convention on Biological Diversity, governments of the world have committed to protecting 17 percent of terrestrial and inland water areas and 10 percent of coastal and marine areas by 2020.

Additionally, forests covered 41.6 million square kilometers (16 million square miles) of the globe in 1990, falling to 40.2 million square kilometers (15.5 million square miles) in 2011, about one-third of the world's land area. The encouraging news is that the annual global deforestation rate decelerated from an average of 0.18 percent in during the 1990s to an average of 0.11 percent in the last decade.

Still Time Enough to Save Ocean Biodiversity

A January 2015 article in *Science* by a team of researchers led by University of California at Santa Barbara ecologist Douglas McCauley seeks to analyze what is happening to marine biodiversity. The good news is that extinctions in the seas appear to have been much rarer than on land. For example, while the IUCN reports that 514 terrestrial animal species have gone extinct since 1500, only 15 marine species have. These include the great auk, Steller's sea cow, and the Caribbean monk seal.

Nevertheless human pressure, especially overfishing, has dramatically reduced the numbers and ranges of many marine species. The study notes, "Aggregated population trend data suggest that in the last four decades, marine vertebrates (fish, seabirds, sea turtles and marine mammals) have declined in abundance by on average 22%. Marine fishes have declined in aggregate by 38%, and certain baleen whales by 80% to 90%." Scripps Institution of Oceanography researcher Jeremy Jackson reported in 2008 that the populations of large open ocean predators such as tuna and sharks have been reduced by 90 percent, oysters by 91 percent, and North Atlantic cod by 96 percent. Nevertheless, as the *Science* article points out, "Marine defaunation, however, has not caused many global extinctions of large-bodied species. Most large-bodied marine animal species still exist somewhere in the ocean."

The chief cause for declining marine populations is overfishing, but habitat degradation could play a bigger role over the course of this century if global warming and ocean acidification continue apace. To rein in excessive exploitation of wild marine populations, the authors recommend among other policies the adoption of incentive-based fisheries. One of the main ways to achieve this is to close open-access fisheries by privatizing them. A 2008 study in *Science* found that implementing such a policy "halts, and even reverses, the global trend toward widespread collapse."

McCauley and his colleagues observe, "Wildlife populations in the oceans have been badly damaged by human activity. Nevertheless, marine fauna generally are in better condition than terrestrial fauna: Fewer marine animal extinctions have occurred; many geographic ranges have shrunk less; and numerous ocean ecosystems remain more wild than terrestrial ecosystems." As a result, the researchers conclude that while the need for action is urgent, there is still time to rescue and restore the biodiversity of the oceans.

Cities Spare Nature

Another extremely positive megatrend with regard to protecting and restoring nature is urbanization. In his 2010 article "How Slums Can Save the Planet," prominent environmental thinker Stewart Brand cited architect Peter Calthorpe's 1985 assertion that "[t]he city is the most environmentally benign form of human settlement. Each city dweller consumes less land, less energy, less water, and produces less pollution than his counterpart in settlements of lower densities." By 2010, the majority of people lived in cities for the first time in history. Demographers expect that 80 percent of people will live in urban areas by 2050 or so. Setting aside the demographic fact that people who live in cities have fewer children, what this trend means is that a lot fewer people will be living on the landscape in the future. Today, about half of the world's population of 7.2 billion people lives in rural areas. Assuming that world population grows to 9 billion by 2050 and that 80 percent do live in cities, that would mean that only 1.8 billion would be on the landscape, as compared to 3.6 billion today. If world population tops out at 8 billion, then only 1.6 billion people would live in the countryside—2 billion fewer people than live there now.

In *The Communist Manifesto*, Karl Marx asserted that bourgeois capitalism fueled the growth of cities and "thus rescued a considerable part of the population from the idiocy of rural life." History has shown that people prefer the opportunities and excitement of city life to rural "idiocy." And the former country dwellers are voting massively in favor of urban living with their feet. Some 60 million people are leaving the countryside to move into cities annually. While some portion may be pushed by war or drought or poverty into cities, most people today are pulled in by the prospect of reinventing themselves, escaping from the narrow strictures of family, class, and community, and a shot at really making it.

As humanity has urbanized, people have become ever less subject to nature's vagaries. For instance, a globally interconnected world made possible by the transportation networks between cities means that a crop failure in one place can be overcome by food imports from areas with bumper crops. Similarly, resources of all types can be shifted quickly to ameliorate human emergencies caused by the random acts of a brutal insensate nature.

Today cities occupy just 2 percent of the earth's surface, but that will likely grow to 3 percent over the next half century. Oddly, environmentalist gadfly Jeremy Rifkin has proclaimed, "In the next phase of human history, we will need to find a way to reintegrate ourselves into the rest of the living Earth if we are to preserve our own species and conserve the planet for our fellow creatures." Actually, he's got it completely backward. Humanity must not reintegrate into nature—in that way lies disaster for humanity and nature. Instead we must make ourselves even more autonomous than we already are from her.

Peak Farmland

Considering that agriculture is the most expansive and intensive way in which people transform natural landscapes, the really good news is that the amount of land globally devoted to food production may be falling as population growth slows and agricultural productivity increases. "We believe that projecting conservative values for population, affluence, consumers, and technology shows humanity peaking in the use of farmland," conclude Jesse Ausubel, the director of the Program for the Human Environment at Rockefeller University, and his colleagues in their 2013 article "Peak Farmland and the Prospect for Land Sparing." They add, "Global arable land and permanent crops spanned 1,371 million hectares in 1961 and 1,533 million hectares in 2009, and we project a return to 1,385 million hectares in 2060." As a result of these trends, humanity will likely restore at least 146 million hectares, an area two and a half times that of France or the size of ten Iowas, and possibly much more land. "Another 50 years from now, the Green Revolution may be recalled not only for the global diffusion of high-yield cultivation practices for many crops, but as the herald of peak farmland and the restoration of vast acreages of Nature," write the researchers. "Now we are confident that we stand on the peak of cropland use, gazing at a wide expanse of land that will be spared for Nature."

The Return of the Forests?

As a consequence of peak farmland, the forests that harbor many species are regrowing around the world. This was confirmed by researchers in a 2006 article on forest trends in the *Proceedings of the National Academy of Sciences* that found that "[a]mong 50 nations with extensive forests reported in the Food and Agriculture Organization's comprehensive *Global Forest Resources Assessment 2005*, no nation where annual per capita gross domestic product exceeded $4,600 had a negative rate of growing stock change." Consider, for example, that between 1960 and 2000, India added 15 million hectares to its forests, an area larger than the state of Iowa. In fact, leaving aside Brazil and Indonesia, forests around the world have increased by about 2 percent since 1990, according to researchers at Resources for the Future.

In 2014, the Food and Agriculture Organization (FAO) reported somewhat less rosy trends with regard to global forest cover. Based on analyzing satellite imagery, the FAO's Global Forest Resources Assessment team concluded that the total forest area in 2010 was 3.89 billion hectares (15 million square miles), which is around 30 percent of the global land area. Between 1990 and 2010, there was a net reduction in the global forest area of around 5.3 million hectares (20,000 square miles) per year. Net deforestation amounted to roughly 106 million hectares (400,000 square miles), an area more than double the size of California. The FAO reports that the extent of boreal, temperate, and subtropical forest area over the past twenty years has largely remained steady and most of the forestland cover reduction occurred in tropical forests.

A February 2015 study also using satellite imagery published in

Geophysical Research Letters by University of Maryland researchers reported similar but accelerated trends for net tropical deforestation. They looked at trends in thirty-four countries that account for 80 percent of tropical forestlands. During the 1990–2000 period the annual net forest loss was 4 million hectares (15,000 square miles) per year. During the 2000–2010 period, the net forest loss rose to 6.5 million hectares (25,000 square miles) per year. The net tropical deforestation in the past twenty years amounts to about 105 million hectares (400,000 square miles).

There is, however, some good news: in 2009 researchers at the Smithsonian Tropical Research Institute estimated that a quarter to a third of the tropical forests that have been cut down by farmers and loggers are now regenerating. Why is this happening? Because like New England farmers before them, small farmers in tropical countries are moving on to more lucrative lives in towns and cities and secondary forests are now growing on their abandoned fields and pastures. In its *State of the World's Forests 2005* report, the UN Food and Agriculture Organization noted that secondary forests in tropical Africa, America, and Asia in 2002 were estimated at 245 million, 335 million, and 270 million hectares, respectively, for a total of 850 million hectares. In addition, the FAO observed that regenerating forests "contribute to biodiversity conservation by relieving pressure on primary forests, by functioning as corridors for the migration of flora and fauna in fragmented landscapes and by maintaining plant and animal genetic resources."

In 2003, some prominent ecologists pessimistically predicted that it was "doubtful that more than 10% of the tropical forests will be protected, and probably more realistic to think of 5% surviving the next 50 years." Smithsonian Tropical Research Institute scientist Joseph Wright rejects this catastrophist prognostication and reports in a 2010 review article that current forest trends suggest that between 64 and 89 percent of the tropical forests present in 2000 will remain forested in 2050. He also cites data showing that the composition of creatures living in naturally regenerating or secondary tropical forests are quite similar to those found in mature tropical forests. "These comparisons suggest that the conservation value of naturally regenerating tropical forests is potentially large," Wright writes. "Fortunately, a wide range of tropical forest species are able to survive in human-modified landscapes, and new research programs are increasingly focused on management to increase the conservation value of human modified, tropical landscapes."

Since tropical forests are thought to harbor huge amounts of biodiversity, because of forest regrowth the pace of species extinctions will likely be lower than many of the more dire projections suggest. At a 2009 conference on deforestation, Eldredge Bermingham, then director of the Smithsonian Tropical Research Institute in Panama, declared: "It's a question of whether or not the biodiversity crisis has been overhyped." He added, "The increase in secondary forest that we are observing may provide a buffer against extinction. Therefore, the extinction crisis isn't as serious as had been touted."

Other recent research suggests that species extinction rates have been exaggerated. In January 2013, three biologists published a study in *Science* reporting that there is no evidence that extinction rates are as high as many are claiming. "Surprisingly, few species have gone extinct, to our knowledge. Of course, there will have been some species which have disappeared without being recorded, but not many, we think," declared Nigel Stork, deputy head of the Griffith School of Environment. Stork also noted that many of the planet's biological hot spots containing high levels of biodiversity are now protected areas.

As mentioned earlier, the extinction rates projected by computer models are based on the species-area curve derived from the theory of island biogeography. In many ways, researchers have been treating areas of tropical forest surrounded by farmland as though they were isolated "islands" surrounded by water. This practice now appears to be wrong. An important 2014 study published in the journal *Nature* finds that in fact farmland and secondary forests are not very much like actual islands. The study looked at the prevalence of species of bats by comparing landscapes of tropical forests surrounded by coffee plantations and pastures with remnant tropical forests on real islands created when an area was flooded to make a lake as part of the Panama Canal. They discovered that unlike on real islands, bats do not go locally extinct in forests embedded in farming landscapes. In fact, the researchers found that "almost one-half of the common bat species were more prevalent in coffee plantations than in forests and only 5 of the 30 species sampled avoided plantations altogether." Instead of island biogeography, the researchers propose a theory of "countryside biogeography" that takes note of the fact that a lot of species, not just bats, can cope with and even thrive in landscapes encompassing natural and modified areas.

The new tools afforded researchers by countryside biogeography render, according to the authors of the 2014 *Nature* study, "more optimistic predictions of biodiversity loss than classic tools because they incorporate the conservation value of the countryside." One of the authors of the new *Nature* study, Stanford University researcher Chase Mendenhall, agreed, "The current model of nature reserves works toward separating humans and nature. That's good to a point, but it can't be the only approach. Today, the world's biodiversity is living with humans, not apart from them. Increased integration looks like the way forward."

The Myth of Pristine Nature

Indeed, more biologists are developing a greater appreciation for how the nature that exists all around us can and should be conserved. "Nature is almost everywhere. But wherever it is, there is one thing nature is not: pristine," explains science journalist Emma Marris in her engaging 2011 book *Rambunctious Garden: Saving Nature in a Post-Wild World*. She adds, "We must temper our romantic notion of untrammeled wilderness and find room next to it for the more nuanced notion of a global, half-wild rambunctious garden, tended by us." These assertions will distress both environmental activists and many ecologists who are in thrall to the damaging cult of pristine wilderness and the false ideology of the balance of nature. But it should encourage and inspire the rest of us.

University of Maryland ecologist Erle Ellis urges us to think about the biological world as being increasingly composed of anthropogenic biomes, or anthromes. Anthromes are novel ecosystems formed through interactions between human activities and natural landscapes and waters. Anthromes include urban, village, agricultural, and range landscapes. Ellis and his colleagues point out that anthromes are mosaics of land used for agriculture or infrastructure and unused or lightly used areas (e.g., steep slopes, wetlands, woodlots) that typically include remnant ecosystems harboring native species.

Lots of environmentalists abhor the notion that Earth is an extensively modified used planet. They believe that novel human-made ecosystems are "degraded" and they often seek to try to return landscapes back to some preferred ecological baseline condition. "For many conservationists, restoration to a pre-human or a pre-European baseline is seen as healing a wounded or sick nature," explains Marris. "For others, it is an ethical duty. We broke it; therefore we must fix it. Baselines thus typically don't act as a scientific before to compare with an after. They become the good, the goal, the one correct state." But what is so good about historical ecosystems?

Marris argues that this preference for setting and trying to maintain ecological baselines was generated from the cult of pristine wilderness preached by nature romantics like John Muir. Muir is famous for advocating that the Yosemite Valley in California's Sierra Nevada Mountains be turned into a national park. As Marris notes, wild nature for Muir was a necessity for "tired, nerve-shaken, over-civilized people" suffering from "the vice of over-industry and the deadly apathy of luxury." And for some people it might be—but that is not a scientific claim about ecosystems and their "integrity."

In fact, there is precious little scientific support for the ideology that pristine nature is somehow "better" than the mélange that humanity has created by moving species around the globe. For example, Marris visits Hawaii, where half of the plant species now living on the islands are non-native. One brave younger ecologist, Joe Mascaro, studies novel ecosystems that are developing on Hawaii that incorporate both native and non-native species. Among other things, Mascaro "found that the novel forests, on average, had just as many species as native forests" and "that in many measures of forest productivity, such as nutrient cycling and biomass, novel forests matched or outproduced the native forests."

Marris contrasts Mascaro with another ecologist, Christian Giardina, who helps manage the Laupahoehoe Natural Area Reserve in Hawaii, from which he wants to extirpate non-natives. Yet even Giardina muses, "Are we so religious about this biodiversity ethic that we need to be called on it?" He answers his own question: "If you really dig down to why we should care, you end up with nothing. You are running on faith that we should care."

Although Marris doesn't cite him, she is plowing much the same intellectual ground as George Mason University philosopher Mark Sagoff. Sagoff has challenged ecologists to name any specifically ecological criterion by which scientists can objectively determine whether an ecosystem whose history they don't know is inhabited by species that have self-organized and coevolved without human interference or a hodgepodge of introduced species that share no evolutionary history. Ecologists, in fact, cannot objectively distinguish between pristine and hodgepodge.

"Imagine that an alien scientist from outer space were to visit both New Zealand and Great Britain," suggest biologists Dov Sax and James Brown, from Brown University and the University of New Mexico, respectively. "Would this individual be able to distinguish which species are native and exotic, and would it be able to demonstrate that invaders have caused more damage or disruption to ecological processes than natives?" Their answer to both questions is no.

"If there were any but magical thinking behind the idea that ecosystems are complex, adaptive systems, scientists could tell by observation and experiment which biotic components play by the rules and which do not. They could tell which is an ancient complex, adaptive, heirloom ecosystem, and which a recent hodgepodge," observes Sagoff.

"In fact, biologists apparently have no way to tell other than by documenting the history of a site whether it represents (1) an ancient, co-evolved community and its states or (2) an agglomeration of colonizing species gathered in the wake of human activity." Since it is the case that "biologists cannot tell by observation or experiment which system is 'co-evolved' and which 'novel' or, indeed, which species are long-timers and which newcomers, then it becomes a question of faith not science" that pristine ecosystems exhibit some kind of superior integrity to human-influenced ecologies.

What Balance of Nature?

Even worse, the widespread notion of the "balance of nature" turns out to be scientifically specious. Early in the twentieth century influential ecologist Frederic Clements developed the theory that ecological communities act like superorganisms working together through a directional and deterministic process of succession toward a stable climax state. Each participant in the climax ecosystem supposedly is fitted tightly into niches as a result of coevolving together. Once achieved, the climax state is exquisitely balanced unless disturbed.

In contrast to these holistic notions, ecologist Henry Gleason, a contemporary of Clements, developed his "individualistic" hypothesis, arguing that ecosystems were assembled by chance, depending on what species got to the landscape first and were successful in living with other species as they arrived. There was no deterministic climax toward which discrete ecosystems were aiming.

For the most part, twentieth-century ecologists fell into Clements's camp. This concept of nature has had significant consequences for our modern understanding of the human role in the environment. University of Maryland public policy professor Robert Nelson observes that Clements's "theory of the climax state also includes a moment when original sin arrived in the world. Human beings were not part of the original ecological order, as Clements described it, but a foreign element." Just as in the biblical account of Adam and Eve in the Garden of Eden, the human quest for knowledge has disturbed prelapsarian harmony, unleashing evil across the natural world. "Only if human beings renounced their false pride and arrogance and humbly accepted a lesser place within creation, leaving the climax state to evolve undisturbed, would there be any hope for the future," comments Nelson. "So far as practically possible, nature should be left untouched by human hands." Clearly, this conviction informs the prognostications of doom by environmentalist soothsayers such as Rachel Carson, Paul Ehrlich, Lester Brown, Bill McKibben, Jeremy Rifkin, the Club of Rome, and so many others during the past half a century.

Appreciating Novel Ecosystems

Today, most ecologists recognize that Gleason was far more right than Clements. There is no balance of nature and assemblages of plants and animals occur mainly by chance. For example, University of California at Davis researcher Arthur Shapiro asserted in a May 2014 lecture at the Commonwealth Club in San Francisco that most evidence is against the idea of stable interdependent communities as the norm in nature. Wheaton College biologist John Kricher concurs: "As a result of research over the past several decades, ecologists have come to understand the reality of ecosystem dynamics, and have largely abandoned the notion that nature exists in some sort of meaningful natural balance." Plants and animals come together and find conditions that enable them to survive in the same shared space. As evidence, Shapiro cites the research of University of Minnesota ecologist Margaret Davis, who found that pollen core data shows that trees recolonizing lands after glaciation don't move in "communities." The tree species migrate at different rates. Consequently, contemporary northern temperate forests are composed of an assemblage of species that mixed together as they raced northward out of various separate refugia as the glaciers retreated at the end of the last ice age.

Of course, species do interact, not because most of them coevolved into tight relationships, but through ecological fitting. Ecological fitting is the process whereby organisms colonize and persist in novel environments, use novel resources, or form novel associations with other species as a result of the suites of traits that they carry at the time they encounter the novel condition. If species could not take advantage of new resources or locations, then it would be impossible for introduced species to survive in novel environments.

One of the more fascinating novel ecosystems has been created on Ascension Island during the past 150 years. Ascension Island is about as isolated as a piece of land can get, sitting in the Atlantic Ocean about midway between Africa and South America. When the British claimed authority over the uninhabited barren hunk of stone in the early nineteenth century, it was frequently likened to a "cinder" or a "ruinous heap of rocks." The new owners named Ascension's central peak White Mountain, after the color of the bare rocks of which it was composed.

In 1846, botanist Joseph Hooker from the Royal Botanic Gardens at Kew visited Ascension and decided to try transplanting a wide variety of plants onto the island. A century and a half later, the result has been an "accidental rain forest." White Mountain, now renamed Green Mountain, is covered with an extensive cloud forest consisting of guava, banana, wild ginger, bamboo, the Chinese glory bower and Madagascan periwinkle, Norfolk Island pine, and eucalyptus from Australia. Because of the man-made microclimate, what used to be a desert island now features several permanent streams.

Ascension Island undercuts the conventional ecological wisdom that tropical rain forests are supposed to take millions of years to form. Species don't need to coevolve to create fully functioning ecosystems; they make the best of what they have. And what happened on Ascension has been happening all around the world as people have moved thousands of species from their native habitats to new locales, increasing species richness. Wherever human beings have gone in the past two centuries, we have increased local and regional biodiversity.

Yet "the popular view [is] that diversity is decreasing at local scales," the Brown University biologist Dov Sax and the University of California at Santa Barbara biologist Steven Gaines report in a 2003 article for *Trends in Ecology and Evolution*. Sax and his University of New Mexico colleague James Brown pointed out in a 2007 roundtable in *Conservation* that "North America presently has more terrestrial bird and mammal species than when the first Europeans arrived five centuries ago." Sax and Gaines's observations were bolstered by an April 2014 article, "Assemblage Time Series Reveal Biodiversity Change but Not Systematic Loss," published in *Science* by a team of researchers led by University of St. Andrews biologist Maria Dornelas. Dornelas and her colleagues analyzed a massive data set covering more than 35,000 mammal, bird, fish, invertebrate, and plant species from marine, freshwater, and terrestrial biomes ranging from the poles to the tropics. The data comprised a hundred individual time series of species composition. "Surprisingly, we did not detect a consistent negative trend in species richness," reported the researchers. Instead, they found that while new and different species have often moved into any given area, the overall diversity of species was in general not declining.

While some introduced species do outcompete natives and contribute to their extinction, that phenomenon is relatively rare. On the whole, the actual number of species in any given area has tended to increase. For example, New Zealand's 2,000 native plant species have been joined by 2,000 from elsewhere, doubling the plant biodiversity of its islands. Meanwhile, only three species of native plants have gone extinct. In California, an additional 1,000 new species of vascular plants have joined the 6,000 native species in the Golden State, while fewer than 30 species have gone extinct. Similar increases in plant diversity can be seen around the globe.

As noted earlier, species that have become extinct and are most in danger of extinction are those that dwell in isolated habitats such as oceanic islands or freshwater streams. In a 2008 article for the *Proceedings of the National Academy of Sciences*, Sax and Gaines note that thousands of oceanic bird species went extinct as Polynesians spread across the Pacific, bringing not only themselves but also hungry rats.

Nevertheless, they point out, the overall species richness of the plant life on Pacific islands has increased considerably, and bird species richness has remained about the same, since the number of extinctions has been balanced by a number of new species moving in.

The richness of mammalian and freshwater species on Pacific islands has dramatically increased as well; it was nearly impossible for animals like rats, pigs, deer, lizards, frogs, catfish, and trout to colonize islands on their own. In addition, while some freshwater species in continental streams and lakes have gone extinct, most now harbor more species than they did before. Hawaii is, for example, home to more than 2,500 new species of invertebrates.

In many cases, the newcomers may actually benefit the natives. In a 2010 review article in the *Annual Review of Ecology, Evolution, and Systematics,* the Rutgers ecologist Joan Ehrenfeld reported that rapidly accumulating evidence from many introduced species of plants and animals shows that they improve ecosystem functioning by increasing local biomass and speeding up the recycling of nutrients and energy. For example, zebra mussels are very effective filter feeders that have helped clear up the polluted waters of the Great Lakes enough to permit native lake grasses and other plants to flourish.

A 2012 review article in *Trends in Ecology and Evolution* surveying the literature on the effects of introduced species on ecosystem functioning reported that a "meta-analysis of over 1000 field studies showing that, although regional native species richness has often declined, primary production and several ecosystem processes were usually maintained or enhanced as a result of species introductions." The researchers further conclude, "What is clear is that ecological theory does not automatically imply that a global decline in species richness will result in impaired functioning of the world's ecosystems." In a 2003 *Science* article, "Prospects for Biodiversity," United Nations Environment Programme researcher Martin Jenkins noted, "In truth, ecologists and conservationists have struggled to demonstrate the increased material benefits to humans of 'intact' wild systems over largely anthropogenic ones [like farms]. . . . Where increased benefits of natural systems have been shown, they are usually marginal and local." Jenkins added that even if the dire projections of global extinction rates being made by conservation advocates are correct, they "will not, in themselves, threaten the survival of humans as a species."

Only when the ecologically correct ideologies that blind us are up-ended can we can see the real nature that is all around us. Baselines are properly transformed into aesthetic choices rather than "scientific" mandates. Consider the ambitious Pleistocene Rewilding proposal in which proxy wild species from Africa might be used to replace those North American species killed off by early peoples. African cheetahs might chase after pronghorns, and elephants graze where mastodons once roamed.

A small version of rewilding is the fascinating Oostvaardersplas-sen experiment in the Netherlands, where researchers are designing an ecosystem that aims to mimic what Northern Europe might have looked like 10,000 years ago. It is stocked with herds of Konik horses and Heck cattle, thought to be respectively similar to the Tarpan horses and the aurochs that once roamed Europe. The newly constructed ecosystem has attracted many wild species that have long been absent from the Netherlands. It is still missing predators, but wolves are apparently mov-ing westward from Eastern Europe. "The Paleolithic landscape at the Oostvaardersplassen is a human creation, the result of human interven-tion, and will require continual human management and maintenance, as does any ecological restoration," points out Sagoff.

Marris and Sagoff are correct that the conservation and apprecia-tion of nature can take place at far less exotic locations, such as back-yards, city parks, farms, and even parking lots. If biodiversity is what is of interest, Marris notes that the Los Angeles area is home to 60 na-tive tree species, but now hosts 145 species. "With eight to eleven tree species per hectare, L.A. is more diverse than many ecosystem types," Marris writes. Another researcher has identified 227 species of bee liv-ing in New York City. And if some of us choose to conserve some areas as "pristine" with regard to some preferred aesthetic baseline, that's okay. Certainly science can be used to help achieve that goal, but such areas become in effect wilderness gardens, maintained, as Marris observes, by "perpetual weeding and perpetual watching."

Since there is no goal or end state toward which any particular ecosystem is heading, who is to say that landscapes and ecosystems modified by human activities are somehow inferior, sick even, and in need of healing? In his 2001 *BioScience* article, "Values, Policy, and Ecosystem Health," Robert Lackey, a fisheries biologist at Oregon State University, pointed out that "ecosystems have no preferences about their states." How do we know whether or not an acre of land would "prefer" to be a swamp or a cornfield? As Lackey notes, either of them could be considered "healthy" depending on what human preferences are being implemented. "To a conservationist interested mainly in biodiversity, we have degraded nature, but to an agronomist, we have altered wild land to make it better serve humans," noted the Nature Conservancy's Peter Kareiva and his colleagues in their 2007 *Science* article "Domesticated Nature: Shaping Landscapes and Ecosystems for Human Welfare."

"Humans must proactively manage ecosystems based on such carefully considered goals as selective conservation of threatened species, maximization of local and regional biodiversity, maintenance of watersheds and soil systems, and other essential functions provided by natural ecosystems," argued Wheaton College researcher John Kricher. "In the next millennium, the balance of nature is what humanity will make it to be." It would be well if more ecologists and environmental activists took to heart Marris's chief insight about conservation: "There is no one best goal." She bravely and correctly concludes, "We've forever altered the Earth, and so now we cannot abandon it to a random fate. It is our duty to manage it. Luckily, it can be a pleasant, even joyful task if we embrace it in the right spirit. Let the rambunctious gardening begin."

CONCLUSION:
ENVIRONMENTAL RENEWAL IN
THE TWENTY-FIRST CENTURY

THE END OF THE WORLD IS NOT NIGH. Far from it. Humanity does face big environmental challenges over the course of the coming century, but the bulk of the scientific and economic evidence shows that most of the trends are positive or can be turned in a positive direction by further enhancing human ingenuity. Let's briefly review that evidence and those trends.

Human population growth is slowing and will very likely peak at around 8 to 9 billion in this century and begin falling. This virtuous trajectory is the result of a combination of happy developments, not the least of which is expanding education and oppportunities for girls and women around the world. The process of economic growth reduces child mortality, which in turn encourages parents to have fewer children and invest more in their health and education. Increasing agricultural productivity ameliorates hunger and liberates people from the fields to seek better opportunities in cities. Rising agricultural yields that result from the application and spread of modern farming technologies, most especially including biotechnology, also means that the amount of land devoted to crops and pasturage is shrinking. Humanity has already likely reached "peak farmland," which means that huge swaths of land will be restored to wild nature over the course of this century. As people abandon the landscape, this greatly enhances the prospects for protecting and preserving the planet's biodiversity.

People at the end of the century will be much wealthier than we are today. Consider again the "middle-of-the-road" projection for economic growth over the rest of the century done by the Organisation for Economic Development and Co-operation's Environment Directorate. In that nothing-special scenario, world population will have peaked at around 9.6 billion in 2065 and fallen to just over 9 billion by 2100. The world's economy will have grown more than eightfold from $67 trillion to $577 trillion (2005 dollars). Average income per person globally will have increased from around $10,000 today to $60,000 by 2100. US annual incomes would average just over $100,000. That amount of wealth enables people to buy a lot in the way of health, education, and wild nature.

Advances in human ingenuity can and in many instances are already beginning to reduce the deleterious side effects of resource usage. The evidence shows that economic growth ultimately results in a cleaner environment. For example, as the developed economies grew over the past four decades, their levels of air pollution steeply declined. Whatever retards economic growth will also retard environmental improvement. Ominous forecasts that humanity is about to run out of vital resources have proven false. Over the course of the last couple of centuries, the availability of resources has increased and the prices of resources have fallen due to technological progress and expansion of free markets. Markets also drive industries and consumers to cut their costs by using resources ever more efficiently. This ceaseless drive for efficiency is already resulting in the dematerialization of some aspects of the economy and the further withdrawal of humanity from nature.

Fears about the effects of modern technologies have turned out to be greatly exaggerated. Widespread predictions of vast cancer epidemics or a rising tide of infertility due to falling sperm counts as a result of exposures to trace amounts of synthetic chemicals have been disproven. Instead of a rising epidemic of tumors, cancer incidence rates have been falling in the United States for the past two decades. The activist insistence on increased precaution toward adopting new technologies very often results in leaving in place more dangerous and polluting old technologies.

The attack on biotech-enhanced crops is one of the more egregiously dishonest activist campaigns. Every independent scientific organization that has evaluated modern biotech crops has found them safe to eat and safe for the environment. To repeat: No one has gotten so much as a cough, sniffle, sneeze, or stomachache from eating foods made with ingredients from modern biotech crops. Among other benefits, biotech crops have reduced pesticide usage, boosted crop productivity, and improved the incomes of farmers around the world.

Man-made climate change is a problem, but it does not portend the end of the world. The solution to future climate change is the same as the remedy for other environmental problems—the application of human ingenuity and technology. Progressives want to use the "climate crisis" as a stalking horse to scare the rest of us into adopting their vast schemes to transform the world's economy into some kind of post-capitalist utopia. Consequently, progressives strongly support United Nations negotiations aiming at a global treaty that would impose central planning on the climate. So far, those negotiations have utterly failed.

It is unfortunately the case that government meddling on a global scale has massively distorted energy markets through pervasive subsidies, mandates, and price controls. The result is retarded innovation in the technologies of energy generation. A big first step toward renovating our energy supply systems would be to eliminate those impediments to understanding the real comparative benefits and costs of the production and use of energy. Ultimately, the better and far more effective way to ameliorate and avert future climate change is to mobilize human ingenuity through market processes to drive down the costs of no-carbon energy sources. Despite the constraints on innovation caused by government interference, notable advances in no-carbon energy generation technologies have already been made, ranging from innovative nuclear reactor designs to more efficient and cheaper solar panels.

New technologies and wealth produced by human creativity will spark a vast environmental renewal in this century. Most global trends suggest that by the end of this century, the world will be populated with fewer and much wealthier people living mostly in cities fueled by cheap no-carbon energy sources. As the amount of land and sea needed to supply human needs decreases, both cities and wild nature will expand, with nature occupying or reoccupying the bulk of the land and sea freed up by human ingenuity. Nature will become chiefly an arena for human pleasure and instruction, not a source of raw materials. I don't fear for future generations; instead, I rejoice for them.

NOTES

Introduction

xv   *"To end our long, self-imposed exile":* Jeremy Rifkin, *Algeny: A New Word—A New World.* New York: Penguin, 1984, 252.

xv   *Vaclav Smil calculates that:* Vaclav Smil, *Making the Modern World: Materials and Dematerialization.* New York: Wiley, 2013.

xvi   *the Antarctic ozone shows a:* J. Kuttippurath et al., "Antarctic Ozone Loss in 1979–2010: First Sign Of Ozone Recovery." *Atmospheric Chemistry and Physics* 13 (2013): 1625–1635. www.atmos-chem-phys.net/13/1625/2013/acp-13-1625-2013.pdf.

xvi   *"The world is in transition":* Lester Brown, "New Era of Food Scarcity Echoes Collapsed Civilizations." February 7, 2013, www.earth-policy.org/book_bytes/2013/fpepch1.

xvi   *"What are the chances":* Paul Ehrlich cited by Kevin Kelley, "UVM Audience Warned of Looming Global Collapse," *Seven Days Vermont,*

May 1, 2013. www.sevendaysvt.com/OffMessage/archives/2013/05/01/
uvm-audience-warned-of-looming-global-collapse.

xvi *global oil production had peaked:* Jörg Schindler, "Peak Oil Could
Trigger Meltdown of Society." Energy Watch Group, October 23, 2007.
www.yubanet.com/artman/publish/article_68545.shtml.

xvi *"The world is at, nearing":* Richard Heinberg, "Beyond the Limits
to Growth." Post Carbon Institute, 2010. www.garfieldfoundation.org/
resources/docs/PCReader-Heinberg-Limits.pdf.

xvi *"It could be a scary future":* Center for Biological Diversity, "The
Extinction Crisis," 2014. www.biologicaldiversity.org/programs/
biodiversity/elements_of_biodiversity/extinction_crisis/.

xvi *"Warming of the climate system":* Thomas Stocker and Qin Dahe,
"Overview of the IPCC WGI Report." *Climate Change 2013: The
Physical Science Basis.* Intergovernmental Panel on Climate Change,
September 2013. www.ipcc.ch/pdf/unfccc/cop19/1_stocker13sbsta.pdf.

xviii *"holds that individuals can be expected":* Dan M. Kahan et al.,
"The Tragedy of the Risk-Perception Commons: Culture Conflict,
Rationality Conflict, and Climate Change" (2011). Temple University
Legal Studies Research Paper No. 2011-26; Cultural Cognition Project
Working Paper No. 89; Yale Law and Economics Research Paper No.
435; Yale Law School, Public Law Working Paper No. 230. Available
at SSRN: ssrn.com/abstract=1871503 or dx.doi.org/10.2139/ssrn.1871503.

1. Peak Population?

2 *In premodern societies:* Tony Volk and Jeremy Atkinson, "Is Child
Death the Crucible of Human Evolution?" *Journal of Social, Evolutionary, and Cultural Psychology* 2.4 (2008): 247. shell.newpaltz.edu/jscc/
articles/volume2/issue4/NEEPSvolkatkinson.pdf.

2 *Global average life expectancy:* Haidong Wang et al., "Age-Specific
and Sex-Specific Mortality in 187 Countries, 1970–2010: A Systematic
Analysis for the Global Burden of Disease Study 2010." *Lancet*
380.9859 (2013): 2071–2094.

3 *The upshot will be:* Wolfgang Lutz, William Butz, and K. C. Samir,
eds., *World Population and Human Capital in the 21st Century,* "Chapter 10: The Rise of Human Capital and the End of World Population

Growth," 519, Oxford University Press, 2014; and K. C. Samir et al., "Results of New Wittgenstein Centre Population Projections by Age, Sex and Level of Education for 171 Countries," Joint Eurostat/UNECE Work Session on Demographic Projections, October 29–31, 2013. staging .unece.org/fileadmin/DAM/stats/documents/ece/ces/ge.11/2013/WP _17.2_01.pdf.

3 *A year earlier:* William Paddock and Paul Paddock, *Famine 1975! America's Decision: Who Will Survive?* Boston: Little, Brown, 1967.

4 **Back in 1963:** Paul Diehl and Nils Petter Gleditsch, *Environmental Conflict.* Boulder, CO: Westview, 2001.

4 **In 1967, Brown:** Lester Brown, "The World Outlook for Conventional Agriculture," *Science,* November 3, 1967, 604.

4 **In 1974, Brown:** Lester Brown, "Global Food Insecurity," *The Futurist* 8.2 (1974), 56.

4 **In 1989, Brown:** Lester Brown, "Feeding Six Billion," *World Watch,* September/October 1989, 32.

4 **Brown contended:** Brown cited in Vaclav Smil, *Feeding the World: A Challenge for the 21st Century.* Boston: MIT Press, 2002, 12.

4 **In 1995, Brown:** Lester Brown, "Facing Food Scarcity," *World Watch,* November/December 1995, www.worldwatch.org/node/407.

4 **In 1996, Brown:** Lester Brown, *Tough Choices: Facing the Challenge of Food Scarcity.* New York: Norton, 1996. www.ecobooks.com/books/ tough.htm.

4 **In a 2012** Scientific American*:* Lester Brown, "Could Food Shortages Bring Down Civilization?" *Scientific American,* May 2009. www .scientificamerican.com/article/civilization-food-shortages.

4 **"The world is in transition":** Lester Brown, "New Era of Food Scarcity Echoes Collapsed Civilizations." February 7, 2013, www.earth -policy.org/book_bytes/2013/fpepch1.

5 **"Dr. Paul Ehrlich says":** *New York Times,* November 25, 1969, 19. From Nexis abstract.

7 **During a May 2013:** Tim Johnson, "Are We Doomed? Probably, but Maybe Not." *Burlington Free Press,* May 2, 2013.

8 **"The world faces":** David Pimentel, "World Overpopulation." *Environment, Development and Sustainability* 14.2 (2012): 151–152. www.skil .org/Qxtras_folder-2/david_pimenteleditorial.pdf.

8 *"The world's biggest problem?"*: Mary Ellen Harte and Anne Ehrlich, "The World's Biggest Problem? Too Many People," *Los Angeles Times,* July 21, 2011. articles.latimes.com/2011/jul/21/opinion/la-oe-harte-population-20110721.

8 *We can choose to limit:* Alan Weisman, *Countdown: Our Last, Best Hope for a Future on Earth?* Boston: Little, Brown, 2013, 40.

9 *"To ecologists who"*: Paul Ehrlich and Anne Ehrlich, *The Population Explosion.* New York: Simon & Schuster, 1990, 34.

9 *"The problem of human"*: Russell Hopfenberg, "Human Carrying Capacity Is Determined by Food Availability." *Population and Environment* 25.2 (November 2003): 1–7. panearth.org/WVPI/Papers/Carrying Capacity.pdf.

10 *Of the nineteen countries:* Population Reference Bureau, 2013 World Population Data Sheet.

12 *Ehrlich doubled down:* Paul Ehrlich, "Eco-Catastrophe." *Ramparts,* 1969, 25.

12 *Both Pakistan and India:* *Economic Times of India,* "India Likely to Export 18 Million Tonnes Rice, Wheat in 2013/14: Report," February 24, 2014. articles.economictimes.indiatimes.com/2014-02-24/news/47635825_1_top-rice-exporter-global-wheat-prices-government-warehouses.

13 *The FAO . . . reports:* FAOSTAT. faostat3.fao.org/faostat-gateway/go/to/download/O/OA/E.

13 *per capita consumption:* Mette Wik, Prabhu Pingali, and Sumiter Broca, "Global Agricultural Performance: Past Trends and Future Prospects," background paper for the World Development Report 2008, World Bank. siteresources.worldbank.org/INTWDR2008/Resources/2795087-1191427986785/Pingali-Global_Agricultural_Performance.pdf.

14 *"plant breeding efforts"*: Keith W. Jaggard, Aiming Qi, and Eric S. Ober, "Possible Changes to Arable Crop Yields by 2050." *Philosophical Transactions of the Royal Society B* 365 (August 16, 2010): 2835–2851. rstb.royalsocietypublishing.org/content/365/1554/2835.full.

14 **"if during the next"**: Paul Waggoner, "How Much Land Can Ten Billion Spare for Nature?" Jesse H. Ausubel and H. Dale Langford, eds., *Technological Trajectories and the Human Environment,* National

Academy of Engineering, 1997, 56–73. www.nap.edu/openbook.php ?record_id=4767&page=56.

14 *India produces 31 bushels:* Ronald Phillips, "Mobilizing Science to Break Yield Barriers." Background paper to the CGIAR 2009 Science Forum workshop: "Beyond the Yield Curve: Exerting the Power of Genetics, Genomics and Synthetic Biology," "2009, 17. www.scienceforum 2009.nl/Portals/11/BGWS4.pdf.

15 *that past population growth:* Julio A. Gonzalo, Félix-Fernando Muñoz, David J. Santos, "Using a Rate Equations Approach to Model World Population Trends." *Simulation: Transactions of the Society for Modeling and Simulation International* 89 (February 2013): 192–198.

16 *"Overpopulation was a spectre":* "A Model Predicts That the World's Populations Will Stop Growing in 2050." *Phys.org.* 4 (April 2013). phys .org/news/2013-04-world-populations.html.

16 *"there is around":* Wolfgang Lutz, Warren Sanderson, and Sergei Scherbov, "The End of World Population Growth." *Nature* 412 (August 2, 2001): 543–545.

16 *"most existing world":* Stuart Basten, Wolfgang Lutz, and Sergei Scherbov, "Very Long Range Global Population Scenarios to 2300 and the Implications of Sustained Low Fertility." *Demographic Research* 28 (May 30, 2013): 1145–1166.

16 *In another 2013 study:* K. C. Samir et al., "Results of New Wittgenstein Centre Population Projections by Age, Sex and Level of Education for 171 Countries." Joint Eurostat/UNECE Work Session on Demographic Projections, October 29–31, 2013. staging.unece.org/fileadmin/DAM/ stats/documents/ece/ces/ge.11/2013/WP_17.2_01.pdf.

16 *projections are way too high:* Sanjeev Sanjay, "Predictions of a Rogue Demographer." *The Wide Angle*, Deutsche Bank, September 9, 2013.

17 *world population stabilization:* Patrick Gerland et al., "World Population Stabilization Unlikely This Century." *Science* 346.6206 (September 18, 2014): 234–237. www.sciencemag.org/content/346/6206/234 .abstract.

17 *In their November 2014 study:* Wolfgang Lutz, William P. Butz, and Samir, K. C. eds., *2014 World Population and Global Human Capital in the Twenty-First Century.* Oxford: Oxford University Press, 2014. webarchive.iiasa.ac.at/Admin/PUB/Documents/XO-14-031.pdf.

19 *This approach suggests:* Bobbi S. Low et al., "Influences on Women's Reproductive Lives: Unexpected Ecological Underpinnings." *Cross-Cultural Research* 42.3 (2008): 201–219.

19 *Another study in 2013:* Bobbi S. Low et al., "Life Expectancy, Fertility, and Women's Lives: A Life-History Perspective." *Cross-Cultural Research* 47.2 (2013): 198–225.

21 *University of Connecticut anthropologists:* Nicola L. Bulled and Richard Sosis, "Examining the Relationship Between Life Expectancy, Reproduction, and Educational Attainment." *Human Nature* 21.3 (2010): 269–289.

22 *US fertility rates:* Michael Haines, "Fertility and Mortality in the United States." EH.Net Encyclopedia, Robert Whaples, ed., March 19, 2008. eh.net/encyclopedia/fertility-and-mortality-in-the-united-states.

22 *"further increases in the rate":* Oded Galor, "The Demographic Transition: Causes and Consequences." Discussion Paper series, 2012, *Forschungsinstitut zur Zukunft der Arbeit,* No. 6334. www.econstor.eu/bitstream/10419/58611/1/715373668.pdf.

23 *OECD economist:* Fabrice Murtin, "On the Demographic Transition 1870–2000." Paris School of Economics Working Paper (2009).

24 *Lutz calculates that:* Wolfgang Lutz, "Toward a 21st Century Population Policy Paradigm." International Institute for Applied Systems Analysis, February 2, 2013.

24 *As Galor noted:* World Trade Organization: www.wto.org/english/res_e/statis_e/its2002_e/chp_2_e.pdf.

24 *as global fertility declined:* John A. Doces, "Globalization and Population: International Trade and the Demographic Transition." *International Interactions* 37 (May 2011): 127–146. www.tandfonline.com/doi/abs/10.1080/03050629.2011.568838#preview.

24–25 *"increasing international exchange":* Mark M. Gray, Miki Caul Kittilson, and Wayne Sandholtz, "Women and Globalization: A Study of 180 Countries, 1975–2000." *International Organization* 60 (Spring 2006): 293–333. www.socsci.uci.edu/~wsandhol/Gray-Kittilson-Sandholtz-IO-2006.pdf.

25 *conclusion is further bolstered:* Ulla Lehmijoki and Tapio Palokangas,

"Population Growth Overshooting and Trade in Developing Countries." University of Helsinki Discussion Paper No. 621, December 7, 2005. ethesis.helsinki.fi/julkaisut/val/kansa/disc/621/populati.pdf

25 *"Fertility rate is highest"*: Seth Norton, "Population Growth, Economic Freedom, and the Rule of Law." PERC Policy Series, February 2002. perc.org/sites/default/files/ps24.pdf.

26 *2013 Index of Economic Freedom: Global Finance,* "Economic Freedom by Country—2013 Ranking." www.gfmag.com/component/content/article/119-economic-data/12450-economic-freedom-by-countryhtml.html#axzz2wzLqcuHl.

27 *Of the 231 countries:* CIA *Factbook* 2013. www.cia.gov/library/publications/the-world-factbook/rankorder/2127rank.html?countryName=Singapore&countryCode=sn&regionCode=eas&rank=224.

28 *continuing global abatement:* Center for Systemic Peace, "Global Conflict Trends," October 14, 2013. www.systemicpeace.org/conflict.htm.

28 *"after the year 2000":* David T. Burbach and Christopher J. Fettweis, "The Coming Stability? The Decline of Warfare in Africa and Implications for International Security." *Contemporary Security Policy* 35.3 (October 2014): 421–445. www.tandfonline.com/doi/abs/10.1080/13523260.2014.963967#.VMbgfcYfku0.

29 *Africa is rapidly urbanizing:* African Development Bank Group, "Urbanization in Africa," December 12, 2012. www.afdb.org/en/blogs/afdb-championing-inclusive-growth-across-africa/post/urbanization-in-africa-10143/.

29 *African urbanites:* David Shapiro and B. Oleko Tambashe, "Fertility Transition in Urban and Rural Areas of Sub-Saharan Africa." Originally presented at the 1999 Chaire Quetelet Symposium in Demography at the Catholic University of Louvain, September 2000. grizzly.la.psu.edu/~dshapiro/Chaire_Quetelet_paper.pdf.

29 *greater urban-rural differential:* Fanaye Tadesse and Derek Headey, "Urbanization and Fertility Rates in Ethiopia." Ethiopia Strategy Support Program II, Working Paper 35, International Food Policy Research Institute. www.ifpri.org/sites/default/files/publications/esspwp35.pdf.

29 *"Would your life be better off?"* Ramez Naam, The Infinite Resource: The Power of Ideas on a Finite Planet. *Lebanon, NH: University Press of New England, 2013.*

2. Is the World Running on Empty?

34 *"recent trends in price":* John Young, "Mining the Earth," Worldwatch Paper 109. Washington, DC: Worldwatch Institute, July 1992, 6.

35 *"real commodity prices":* Office of the Chairman of the Group of 77, "Overview of the Situation of Commodities in Developing Countries," March 2005, www.g77.org/ifcc11/docs/doc-04-ifcc11.pdf.

35 *average prices for energy:* John Bluedorn et al., International Monetary Fund, *World Economic Outlook: Growth Resuming, Dangers Remain,* Chapter 4, "Commodity Price Swings and Commodity Exporters," April 2012, 125.

35 *food price index:* FAO Food Price Index, Food and Agriculture Organization, January, 19 2014. www.fao.org/worldfoodsituation/foodpricesindex/en/.

35 *"The world is at":* Richard Heinberg, "Beyond the Limits to Growth," Post Carbon Reader Series: Foundation Concepts, 2010, 2. www.postcarbon.org/Reader/PCReader-Heinberg-Limits.pdf.

35–36 *"Government and corporate officials":* Michael Klare, *The Race for What's Left: The Global Scramble for the World's Last Resources.* New York: Macmillan, 2012, 7. michaelklare.com/wordpress/wp-content/uploads/2012/04/Excerpt.pdf.

36 *"The world is in transition":* Lester Brown, *Full Planet, Empty Plates: The New Geopolitics of Food Scarcity.* New York: Norton, 2012, 160. www.earth-policy.org/books/fpep/fpepch1.

36 *"It is a matter of":* Hans W. Singer, "The Distribution of Gains Between Investing and Borrowing Countries." *American Economic Review* 40 (1950): 477.

36 *"the global economy witnesses":* David Jacks, "From Boom to Bust: A Typology of Real Commodity Prices in the Long Run," National Bureau of Economic Research Working Paper 18874, March 2013, 2. www.nber.org/papers/w18874.

37 *increases in commodity prices:* Martin Stuermer, "150 Years of Boom and Bust: What Drives Mineral Commodity Prices?," Munich RePEc Archive Paper 51859, December 4, 2013, 9. mpra.ub.uni-muenchen.de/51859/1/MPRA_paper_51859.pdf.

37 *The super-cycles are:* Martin Stuermer, "150 Years of Boom and Bust: What Drives Mineral Commodity Prices?" Munich RePEc Archive Paper 51859, December 4, 2013. mpra.ub.uni-muenchen.de/51859/1/MPRA_paper_51859.pdf.

38 *"Since 1871, the Economist":* Blake Clayton, "Bad News for Pessimists Everywhere," *Energy, Security, and Climate,* Council on Foreign Relations, March 22, 2013.

38 *many researchers believe:* David Jacks, "From Boom to Bust: A Typology of Real Commodity Prices in the Long Run," National Bureau of Economic Research Working Paper 18874, March 2013, 4. www.nber.org/papers/w18874; and Maria Kolesnikova and Isis Almeida, "Goldman Sees New Commodity Cycle as Shale Oil Spurs U.S. Growth." Bloomberg News, January 13, 2014. www.bloomberg.com/news/2014-01-13/goldman-sees-new-commodity-cycle-as-shale-oil-spurs-u-s-growth.html.

39 *all known oil reserves:* Donella H. Meadows, Dennis Meadows, Jørgen Randers, and William W. Behrens III, *The Limits to Growth.* New York: New American Library, 1972, 66.

40 *2013 now ranks:* Tim McMahon, "Historical Crude Oil Prices: Oil Prices 1946–Present," InflationData.com, March 6, 2014. inflationdata.com/Inflation/Inflation_rate/Historical_Oil_Prices_Table.asp.

40 *When the bet was settled:* John Tierney, "Economic Optimism? Yes, I'll Take That Bet." *New York Times,* December 27, 2010. www.nytimes.com/2010/12/28/science/28tierney.html?_r=0.

41 *world had reached the peak:* Werner Zittel and Jörg Schindler, "Crude Oil: The Supply Outlook." Energy Watch Group, October 2007, 117. www.lighttrailuk.com/pdf/ewg_oilreport_oct_2007.pdf.

41 *an analysis released by:* Steve Geisel, "New 'Super-Spike' Might Mean $200 a Barrel Oil." MarketWatch, *Wall Street Journal,* March 7, 2008. www.marketwatch.com/story/goldman-sachs-raises-possibility-of-200-a-barrel-oil.

42 *In 2014, global oil production:* US Energy Information Administration, "Short-Term Energy Outlook," May 2014. www.eia.gov/forecasts/steo/report/global_oil.cfm.

42 *that to meet demand in 2035:* Richard Newell and Stuart Iler, "The Global Energy Outlook." NBER Working Paper 18967, April 2013.

43 *total world petroleum reserves:* International Energy Agency, *Resources to Reserves 2013.* www.iea.org/Textbase/npsum/resources 2013SUM.pdf.

43 *"Oil is not in short supply":* Leonardo Maugeri, "Oil: The Next Revolution," Discussion Paper 2012-10, Belfer Center for Science and International Affairs, John F. Kennedy School of Government, Harvard University, June 2012.

43 *"Conventional recoverable resources":* IEA, *World Energy Outlook 2011: Are We Entering a Golden Age of Gas?,* 7. www.worldenergyoutlook.org/media/weowebsite/2011/WEO2011_GoldenAgeofGasReport.pdf.

44 *2.6 million oil and gas wells:* Wilderness Society, "Oil and Gas Drilling: Some Key Facts," April 2011. beyondoil.files.wordpress.com/2011/04/drilling-in-america-february-2011.pdf.

44 *(OPEC) wells total:* OPEC, Annual Statistical Bulletin 2012, 27. www.opec.org/opec_web/static_files_Project/media/downloads/publications/ASB2010_2011.pdf.

44 *27 and 30 billion barrels of reserves:* Richard Nehring, *Linking U.S. Oil and Gas Reserve Estimates.* RAND Corporation, September 1984, 3. www.rand.org/content/dam/rand/pubs/notes/2009/N2049.pdf.

44 *US proven oil reserves at 29 billion barrels:* US Energy Information Administration, "U.S. Field Production of Crude Oil," 2014. www.eia.gov/dnav/pet/hist/LeafHandler.ashx?n=PET&s=MCRFPUS1&f=A.

45–46 *"an implied cost":* ExxonMobil, *The Outlook for Energy: A View to 2040,* 2014, 32. usaans1.usaa.com/en/energy/energy-outlook.

46 *carbon dioxide prices:* Rachel Wilson et al., "2012 Carbon Dioxide Price Forecast," Synapse Energy Economics, October 4, 2012. www.synapse-energy.com/sites/default/files/SynapseReport.2012-10.0.2012-CO2-Forecast.A0035.pdf.

46 *Reuters polled twenty:* Claire Milhench and Alexander Winning, "Oil by 2020 to Fall to $80 in Real Terms—Reuters Poll." Reuters, October 30, 2013. www.reuters.com/article/2013/10/30/brent-poll -idUSL5N0IJ3ED20131030.

46 *$128 per barrel:* IEA, World Energy Outlook Factsheet, "How Will Global Energy Markets Evolve to 2035?" www.iea.org/media/files/ WEO2013_factsheets.pdf.

46 *real oil prices:* Bilge Erten and José Antonio Ocampo, "Super-Cycles of Commodity Prices Since the Mid-Nineteenth Century," Initiative for Policy Dialogue Working Paper Series, Columbia University, January 2012, 28.

46 *"The age of 'cheap oil' ":* Leonardo Maugeri, "Oil: The Next Revolution," Discussion Paper 2012-10, Belfer Center for Science and International Affairs, John F. Kennedy School of Government, Harvard University, June 2012, 6.

47 *government-owned oil companies:* Silvana Tordo, Brandon Tracy, and Noora Arfaa, World Bank Working Paper 218, National Oil Companies and Value Creation, Vol. 1, 2011. siteresources.worldbank.org/INTOGMC/ Resources/9780821388310.pdf.

47 *In the wake of the Arab Spring:* Michael L. Ross, "Will Oil Drown the Arab Spring?" *Foreign Affairs*, September/August 2011. www .foreignaffairs.com/articles/68200/michael-l-ross/will-oil-drown-the -arab-spring.

49 *"Scarcity leads to shortage":* David Zetland, "Do Smaller Water Footprints Lead to Bigger Profits?" *The Guardian*, November 22, 2010. www.theguardian.com/sustainable-business/water-footprints-bigger -profits.

49 *"Scarcity and shortage are the same":* David Zetland, *Living with Water Scarcity*. Aguanomics Press, 2014, 100. livingwithwaterscarcity .com/.

49 *"Underpricing":* Summary Human Development Report 2006, *Beyond Scarcity: Power, Poverty and the Global Water Crisis*. New York: UNDP, 2006, 52. hdr.undp.org/sites/default/files/hdr_2006_summary_en.pdf.

49 *Even now, 3.6 billion: Drinking Water.* World Health Organization, 2012. www.who.int/water_sanitation_health/monitoring/water.pdf.

49 *foolish policies:* John Parnell, "World on Course to Run out of Water, Warns Ban Ki-Moon." *The Guardian,* May 23, 2013. www.theguardian .com/environment/2013/may/22/world-run-out-water-ban-ki-moon1.

51 *water privatization:* Fredrik Segerfeldt, *Water for Sale: How Business and the Market Can Resolve the World's Water Crisis.* Washington, DC: Cato Books, 2006, 144; see a short version at www.ein.eu/files/ Segerfeldt_Amigo.pdf.

51 *these initial projects:* Jennifer Franco and Sylvia Kay, "The Global Water Grab: A Primer." Transnational Institute, March 13, 2012. www .tni.org/primer/global-water-grab-primer#countries.

52 *"Barriers to adaptation":* IPCC, *Climate Change 2014: Impacts, Adaptation, and Vulnerability,* April 2014, Chapter 3, 76. ipcc-wg2.gov/AR5/ images/uploads/WGIIAR5-Chap3_FGDall.pdf.

55 *recent alarms:* Tom Philpott, "Are We Heading Toward 'Peak Fertilizer'?" *Mother Jones,* November 28, 2012. www.motherjones.com/tom -philpott/2012/11/are-we-heading-toward-peak-fertilizer.

55 *applying nitrogen, phosphorus:* National Agricultural Center and Hall of Fame, Justus von Liebig. www.aghalloffame.com/hall/liebig .aspx.

55 *"humanity faces a Malthusian trap":* James Elser and Stuart White, "Peak Phosphorus." *Foreign Policy,* April 20, 2010. www.foreignpolicy .com/articles/2010/04/20/peak_phosphorus.

55 *"impending shortage of two fertilizers":* Jeremy Grantham, "Be Persuasive. Be Brave. Be Arrested (If Necessary)." *Nature* 491.7424 (November 14, 2012). www.nature.com/news/be-persuasive-be-brave -be-arrested-if-necessary-1.11796.

56 *development of such nutrient-efficient crops:* Xiurong Wang et al., "Overexpressing AtPAP15 Enhances Phosphorus Efficiency in Soybean." *Plant Physiology* 151.1 (September 2009): 233–240. www .plantphysiol.org/content/151/1/233.abstract.

56 *"insufficient economically recoverable lithium":* William Tahil, "The Trouble with Lithium: The Implications of Future PHEV Production for Lithium Demand." Meridian International Research, January 2007. www.inference.phy.cam.ac.uk/sustainable/refs/nuclear/ TroubleLithium.pdf.

57 *"Even with a rapid":* Paul Gruber et al., "Global Lithium Availabil-

ity: A Constraint for Electric Vehicles?" *Journal of Industrial Ecology* 15.5 (October 2011): 760–775. onlinelibrary.wiley.com/doi/10.1111/j .1530-9290.2011.00359.x/abstract.

57 **on the order of 1 billion:** Cyrus Wadia, Paul Albertus, and Venkat Srinivasan, "Resource Constraints on the Battery Energy Storage Potential for Grid and Transportation Applications." *Journal of Power Sources* 196.3 (February 2011), 1598. cyruswadia.com/prof/Publications_files/ Wadia%20et.al.%20Resource%20Constraints%20on%20Battery%20 Storage.pdf.

57 **hike in the price of some raw material:** Chris Rhodes, "Peak Minerals: Shortage of Rare Earths Metals Threatens Renewable Energy," OilPrice.com, July 30, 2012. oilprice.com/Alternative-Energy/Renewable -Energy/Peak-Minerals-Shortage-of-Rare-Earth-Metals-Threatens -Renewable-Energy.html; Harald Ulrik Sverdrup, Deniz Koca, and Kristin Vala Ragnarsdottir, "Metals, Minerals, Energy, Wealth, Food and Population: Urgent Policy Considerations for a Sustainable Society." *Journal of Environmental Science and Engineering B 1* (January 20, 2013): 499–533. www.davidpublishing.com/davidpublishing/Upfile/ 10/21/2013/2013102168383361.pdf.; and Ugo Bardi and Marco Pagani, "Peak Minerals." *The Oil Drum*, October 15, 2007. www.theoildrum .com/node/3086.

58 **"Prices for most rare":** US Department of Defense, Annual Industrial Capabilities Report to Congress, October 2013, 25. www.acq.osd.mil/ mibp/docs/annual_ind_cap_rpt_to_congress-2013.pdf.

58 **Tesla Motors installs:** European Commission, *Futurium*, "Emerging Alternatives to Rare Earth Elements," 2013. ec.europa.eu/digital-agenda/ futurium/en/content/emerging-alternatives-rare-earth-elements.

59 **"The influence of innovation":** Harry Bloch and David Sapsford, "Innovation, Real Primary Commodity Prices, and the Business Cycles," paper presented at the International Schumpeter Society Conference 2010 on Innovation, Organisation, Sustainability and Crises, Aalborg, June 2010, 10.

60 **presence of an EKC-type relationship:** Bishwa Koirala et al., "Further Investigation of Environmental Kuznets Curve Studies Using Meta-Analysis." *International Journal of Ecological Economics and Statistics* 22.S11 (2011). www.ceserp.com/cp-jour/index.php?journal=ijees&page =article&op=view&path[]=1014.

60 *globally, pollution:* edgar.jrc.ec.europa.eu/news_docs/July%2019_v2 .pdf, Version v4.1of the Emission Database for Global Atmospheric Research (EDGAR), July 2010.

60 *sulfur dioxide emissions:* K. Zilmont, S. J. Smith, and J. Cofala, "The Last Decade of Global Anthropogenic Sulfur Dioxide 2000–2011." *Environmental Research Letters* 8.1 (January 2013). iopscience.iop.org/ 1748-9326/8/1/014003/article.

60 *pollution turning point:* Anil Markandya et al., "Empirical Analysis of National Income and SO_2 Emissions in Selected European Countries." *Environmental and Resource Economics* 35 (2006): 221–257. www.environmental-expert.com/Files/6063/articles/9212/1.pdf.

61 *"If consumers dematerialize":* Jesse H. Ausubel and Paul Waggoner, "Dematerialization: Variety, Caution, and Persistence," *Proceedings of the National Academy of Sciences* 105.35 (September 2, 2008): 12774– 12779. www.pnas.org/content/105/35/12774.full.

61 *modern technology enables:* Vaclav Smil, *Making the Modern World: Materials and Dematerialization.* New York: Wiley, 2013.

61 *the amount of energy:* Ramez Naam, *The Infinite Resource: The Power of Ideas on a Finite Planet.* Lebanon, NH: University Press of New England, 2013.

62 *significant gains in energy productivity:* Alliance Commission on National Energy Efficiency Policy, "History of Energy Efficiency." Alliance to Save Energy, January 2013, 4. www.ase.org/sites/ase.org/files/ resources/Media%20browser/ee_commission_history_report_2-1-13 .pdf.

63 *a realistic simulation:* Daniel J. Fagnant and Kara M. Kockelman, "The Travel and Environmental Implications of Shared Autonomous Vehicles, Using Agent-Based Model Scenarios." *Transportation Research Part C: Emerging Technologies* 40 (March 2014): 1–13. www.sciencedirect .com/science/article/pii/S0968090X13002581.

63 *shared autonomous vehicles:* Lawrence Burns, William Jordan, and Bonnie Scarborough, "Transforming Personal Mobility." The Earth Institute, Columbia University, New York, 2013.

64 *resource consumption trends:* Iddo Wernick and Jesse Ausubel, "Making Nature Useless? Global Resource Trends, Innovation, and Implications for Conservation." Resources for the Future, November

5, 2014. www.rff.org/Events/Pages/Making-Nature-Useless-Global
-Resource-Trends-Innovation-and-Implications-for-Conservation.aspx.

66 *peak farmland:* J. H. Ausubel, I. K. Wernick, and P. E. Waggoner, "Peak
Farmland and the Prospect for Land Sparing." *Population and Devel-
opment Review* 38 (2013): 221–242. phe.rockefeller.edu/docs/PDR
.SUPP%20Final%20Paper.pdf.

66 *release of vast areas of land:* Jesse H. Ausubel, "Peak Farmland,"
lecture on December 18, 2012. Symposium in Honor of the 80th Birthday of Paul Demeny and his retirement as editor of *Population and
Development Review.* phe.rockefeller.edu/docs/Peak%20Farmland%20
18%20Dec%20lecture%20Ausubel%281%29.pdf.

66 *crop plants bioengineered:* Kenrick Vezina, "Nitrogen-Efficient Crops:
The Holy Grail of Agricultural Biotech?" Genetic Literacy Project,
March 18, 2013. www.geneticliteracyproject.org/2013/03/18/nitrogen
-efficient-crops-the-holy-grail-of-agricultural-biotech/#
.UwYouIUxbLc.

66 **cultured meat:** Avy Roy, "Why Meat Grown in Labs Is the Next Logical Step for Food Production." *Next Nature,* June 28, 2013. www
.nextnature.net/2013/08/meat-grown-in-labs-is-the-next-logical-step
-for-food-production/.

66 *"building a progressive food system":* New Harvest, www.new
-harvest.org/.

66 *"is likely to play a significant role":* Banning Garrett, "3D Printing:
New Economic Paradigms and Strategic Shifts." *Global Policy* 4.1
(February 2014). 70–75. onlinelibrary.wiley.com/doi/10.1111/1758-5899
.12119/full.

67 *"Additive manufacturing":* Advanced Manufacturing Office, "Additive Manufacturing: Pursuing the Promise." US Department of Energy,
August 2012.

67 *"Sustainable development":* Gro Harlem Brundtland, *Our Common
Future: Report of the World Commission on Environment and Development,* 1987. www.un-documents.net/our-common-future.pdf.

68 *economic growth proceeded:* Angus Maddison, The Maddison Project, Original Maddison Home Page, January 2013. www.ggdc.net/
maddison/maddison-project/home.htm.

68 *ultimately unsustainable societies:* Douglass C. North, John Joseph

Wallis, and Barry R. Weingast, *Violence and Social Orders: A Conceptual Framework for Interpreting Recorded Human History.* New York: Cambridge University Press, 2009.

68 *concur with the analysis:* Daron Acemoğlu and James Robinson, *Why Nations Fail: The Origins of Power, Prosperity, and Poverty.* New York: Crown Business, 2012.

69 *"Many lines of evidence":* Harvey Weiss and Raymond S. Bradley, "What Drives Societal Collapse?" *The Heat Is Online,* January 26, 2001. www.heatisonline.org/contentserver/objecthandlers/index.cfm?id =3629&method=full.

69 *problem-solving institutions:* Joseph Tainter, *The Collapse of Complex Societies.* New York: Cambridge University Press, 1988.

70 *self-organizing critical systems:* Gregory G. Brunk, "Why Do Societies Collapse?: A Theory Based on Self-Organized Criticality." *Journal of Theoretical Politics* 14.2 (April 2002): 195–230. jtp.sagepub.com/ content/14/2/195.abstract.

70 *the start of World War I:* Thomas Kron and Thomas Grund, "Society as a Self-Organized Critical System," *Cybernetics and Human Knowing* 16.1–2 (January 1, 2009): 65–82. www.soziologie.rwth-aachen.de/global/ show_document.asp?id=aaaaaaaaaackhty.

70 *critical point:* Mancur Olson, *The Rise and Decline of Nations: Economic Growth, Stagflation, and Social Rigidities.* New Haven: Yale University Press, 1982.

71 *capitalist development:* Ronald Bailey, "China Needs the Rule of Law." *Reason,* May 16, 2012. reason.com/archives/2012/05/16/china -needs-the-rule-of-law.

71 *"the current economic model":* Sustainable Societies, Responsive Citizens. Declaration adopted at the sixty-fourth annual Conference of the Department of Public Information for Non-Governmental Organizations, Bonn, Germany, September 3–5, 2011. www.un.org/wcm/ webdav/site/dpingorelations/shared/Final%20Declaration/ BonnEng.pdf.

73 *"Is it realistic":* Lucas Bretschger and Sjak Smulders, "Sustainability and Substitution of Exhaustible Natural Resources: How Resource Prices Affect Long-Term R&D-Investments," in *Environment and Sustainable Development,* I. Sundar, ed. New Delhi: APH Publishing Com-

pany, 2006. www.cer.ethz.ch/wif/wif/resec/research/research_seminar/substitution.hp.pdf.

73 *"Every generation has perceived":* Paul Romer, "Economic Growth," *The Concise Encyclopedia of Economics,* 2nd ed. www.econlib.org/library/Enc/EconomicGrowth.html.

3. Never Do Anything for the First Time

76 *"more sorry than safe":* Jonathan H. Adler, "More Sorry Than Safe: Assessing the Precautionary Priniciple and the Proposed International Biosafety Protocol." *Texas International Law Journal,* Spring 2000, 173–205. scholarlycommons.law.case.edu/cgi/viewcontent.cgi?article=1225&context=faculty_publications.

76 *"When an activity":* Science and Environmental Health Network, Wingspread Consensus Statement on the Precautionary Principle, January 26, 1998. www.sehn.org/wing.html.

77 *"is not an anti-science view":* Chris Mooney, "Unequivocal: Today's Right Is Overwhelmingly More Anti-Science Than Today's Left." *DeSmogBlog,* September 9, 2011. www.desmogblog.com/unequivocal-today-s-right-overwhemingly-more-anti-science-today-s-left.

77 *"Assume that all":* Peter Montague, "The Precautionary Principle in the Real World," Environmental Research Foundation, January 21, 2008. www.rachel.org/lib/pp_def.htm.

78 *"The truth of the matter":* George Annas, cited in Ronald Bailey, "Precautionary Tale." *Reason,* April 1999. reason.com/archives/1999/04/01/precautionary-tale/1.

78 *"If the burden of proof":* Cass Sunstein, "The Paralyzing Principle." *Regulation,* Winter 2002–2003, 32–37. object.cato.org/sites/cato.org/files/serials/files/regulation/2002/12/v25n4-9.pdf.

78 *"The problem is":* Adam Thierer, "Technopanics, Threat Inflation, and the Danger of an Information Technology Precautionary Principle." *Minnesota Journal of Law, Science and Technology,* January 25, 2013.

78 *Minotaurs are notoriously:* Rahim Sameer, "The Opera Novice: The Minotaur by Harrison Birtwistle and David Harsent." *The Telegraph,* January 22, 2013. www.telegraph.co.uk/culture/music/opera/

9818147/The-opera-novice-The-Minotaur-by-Harrison-Birtwistle-and
-David-Harsent.html.

79 *overall green plan:* Steve Breyman, cited in Ronald Bailey, "Precautionary Tale." *Reason,* April 1999. reason.com/archives/1999/04/01/precautionary-tale/1.

79 *"the greatest uncertainty":* Søren Holm and John Harris, "Precautionary Principle Stifles Discovery." *Nature* 400: 398 (July 29, 1999), cited in Gary Marchant et al., Council for Agricultural Science and Technology (CAST), *Impact of the Precautionary Principle on Feeding Current and Future Generations.* Issue Paper 52. CAST, Ames, Iowa, 2013.

80 *"The precautionary principle":* Cass R. Sunstein, "Throwing Precaution to the Wind: Why the 'Safe' Choice Can Be Dangerous." *Boston Globe,* July 13, 2008.

80 *five different common cognitive biases:* Cass R. Sunstein, "The Laws of Fear." University of Chicago Law and Economics, Olin Working Paper No. 128 (June 2001). Available at SSRN: ssrn.com/abstract=274190 or dx.doi.org/10.2139/ssrn.274190.

81 *German government decided:* Michael Bastasch, "CO_2 Emissions Have Increased Since 2011 Despite Germany's $140 Billion Green Energy Plan." *The Daily Caller,* April 9, 2014. dailycaller.com/2014/04/09/germanys-140-billion-green-energy-plan-increased-co2-emissions/#ixzz2yaOKrXJF.

81 *"paralyzing principle":* Sunstein, "The Paralyzing Principle."

81 *modern pesticides:* Gary Marchant et al., Council for Agricultural Science and Technology (CAST), *Impact of the Precautionary Principle on Feeding Current and Future Generations.* Issue Paper 52. CAST, Ames, Iowa, 2013.

82 *time from drug discovery to marketing:* David J. Stewart, Simon N. Whitney, and Razelle Kurzrock, "Equipoise Lost: Ethics, Costs, and the Regulation of Cancer Clinical Research." *Journal of Clinical Oncology* 28.17 (June 10, 2010): 2925–2935. jco.ascopubs.org/content/28/17/2925.abstract.

82 *faster FDA drug approvals:* Tomas Philipson et al., "Assessing the Safety and the Efficacy of the FDA: The Case of the Prescription Drug User Fee Acts." National Bureau of Economic Research Working Paper 11724, October 2005. www.nber.org/papers/w11724.pdf?new_window=1.

82 *FDA announces its approval:* Sam Kazman,"Drug Approvals and Deadly Delays." *Journal of American Physicians and Surgeons,* November 1, 2010, 101–103. www.jpands.org/vol15no4/kazman.pdf.

83 *"there is no clear evidence":* "Artificial Sweeteners and Cancer," National Cancer Institute Fact Sheet, 2009. www.cancer.gov/cancer topics/factsheet/Risk/artificial-sweeteners.

83 *"the current body of evidence":* Commission on Life Sciences, *Possible Health Effects of Exposure to Residential Electric and Magnetic Fields.* Washington, DC: National Academies Press, 1997, 2 [online]. fermat.nap.edu/books/0309054478/html/2.html.

83 *seven-year epidemiological study:* Questions and Answers About the National Cancer Institute/Children's Cancer Group Study Finds Magnetic Fields Do Not Raise Children's Leukemia Risk, NIH Press Release, July 2, 1997. www.nih.gov/news/pr/jul97/ncib-02.htm.

84 *"The EMF controversy":* Robert Park, "Voodoo Science and the Power-Line Panic." *Forbes,* May 15, 2000, 128.

84 *ordinance requiring radiation warning labels:* Ronnie Cohen, "San Francisco Surrenders in Fight Over Cell Phone Warnings." Reuters, May 8, 2013. www.reuters.com/article/2013/05/08/us-usa-sanfrancisco -cellphones-idUSBRE9470I720130508.

84 *"to date there is no evidence":* "Cell Phones and Cancer Risk." National Cancer Institute Fact Sheet, 2013. www.cancer.gov/cancertopics/ factsheet/Risk/cellphones.

85 *"full and inclusive assessments":* Eric Hoffman et al. *The Principles for the Oversight of Synthetic Biology.* Friends of the Earth, March 2012.

85 *"rooted in the precautionary principle":* Eric Hoffman, "Global Coalition Calls for Oversight of Synthetic Biology." Friends of the Earth, March 13, 2012.

85 *"synthetic biology does not":* Presidential Commission for the Study of Bioethical Issues, *New Directions: The Ethics of Synthetic Biology and Emerging Technologies.* December 1, 2010, 124.

86 *"a more comprehensive application":* Georgia Miller, "Who's Afraid of the Precautionary Principle." Friends of the Earth, 2010. nano.foe .org.au/node/186.

86 *"not to judge things":* Frédéric Bastiat, "What Is Seen and Unseen."

Selected Essays in Political Economy, Library of Economics and Liberty. www.econlib.org/library/Bastiat/basEss1.html.

87 *Why is it safer":* Jonathan Adler, "The Problems with Precaution: A Principle Without Principle." *The American,* May 25, 2011.

87 *"The rhetoric works":* Aaron Wildavsky and Adam Wildavsky, "Risk and Safety. *"Concise Encyclopedia of Economics,* 2nd ed. www.econlib .org/library/Enc/RiskandSafety.html.

88 *forbade the chemical manufacturer:* Ronald Bailey, "Brain Drain." *Forbes,* November 27, 1989, 261.

89 *"generic focus on new products":* Gary Marchant et al., Council for Agricultural Science and Technology (CAST). *Impact of the Precautionary Principle on Feeding Current and Future Generations.* Issue Paper 52. CAST, Ames, Iowa, 2013.

91 *"The true key to the timing":* Joel Mokyr, *The Gifts of Athena: Historical Origins of the Knowledge Economy.* Princeton, NJ: Princeton University Press, 2002.

91 *"Liberalism and science":* Timothy Ferris in Michael Shermer, "Democracy's Laboratory: Are Science and Politics Interrelated?" *Scientific American,* September 2010. www.scientificamerican.com/ article.cfm?id=democracys-laboratory.

91 *"Human reason can neither predict":* Friedrich Hayek, *The Constitution of Liberty: The Definitive Edition,* ed. Ronald Hamowy. Chicago: University of Chicago Press, 2011, 94.

91 *Wherever the institutions:* Daron Acemoğlu and James Robinson, *Why Nations Fail: The Origins of Power, Prosperity, and Poverty.* New York: Crown Business, 2012.

93 *freedom means the renunciation:* Friedrich A. Hayek, *The Constitution of Liberty.* Chicago: University of Chicago Press, 1978, 572. nazbol .net/library/authors/Friedrich%20August%20Hayek/Friedrich _Hayek%20-%20The_constitution_of_liberty.pdf.

93 *"Nowhere is freedom":* Hayek, *The Constitution of Liberty,* 394.

4. What Cancer Epidemic?

96 *"growing scientific consensus":* Pesticide Action Network, "Cancer." www.panna.org/your-health/cancer.

96 *"Consider the deadly":* Paul Farrell, "World War III: The 12-Bomb Equation." *Wall Street Journal* MarketWatch, September 29, 2009. mammonmessiah.blogspot.com/2010/06/paul-farrell-world-war-iii-12 -bomb.html.

96 *"the true burden of":* Tiffany O'Callaghan, "President's Panel Analyzes Environmental Cancer Impact." *Time,* May 6, 2010. healthland.time .com/2010/05/06/presidents-panel-analyzes-environmental-cancer -impact/.

96 *"With nearly 80,000":* LaSalle D. Leffall Jr. and Margaret L. Kripke, *Reducing Environmental Cancer Risk: What We Can Do Now.* National Cancer Institute, President's Cancer Panel, April 2010. deainfo.nci.nih .gov/advisory/pcp/annualReports/pcp08-09rpt/PCP_Report_08-09 _508.pdf.

97 *"Another possible threat":* Paul R. Ehrlich and Anne H. Ehrlich, "Can a Collapse of Global Civilization Be Avoided?" *Proceedings of the Royal Society B* 282.1801 (March 7, 2013). rspb.royalsocietypublishing.org/ content/280/1754/20122845.full.

97 *seven out of ten Americans:* Kevin Stein, Luhua Zhao, Corrine Crammer, and Ted Gansler, "Prevalence and Sociodemographic Correlates of Beliefs Regarding Cancer Risks." *Cancer* 100.5 (September 1, 2007): 1139–1148. onlinelibrary.wiley.com/doi/10.1002/cncr.22880/full.

97 *"Because the rate":* Rebecca Viksnins Snowden, "Cancer Death Rate Steadily Declining," citing John R. Seffrin, American Cancer Society, May 27, 2009. www.cancer.org/cancer/news/cancer-death-rate-steadily -declining.

98 *"It is gratifying":* Harold Varmus, cited in "Report to the Nation Finds Continued Declines in Many Cancer Rates." National Cancer Institute, March 21, 2011. www.cancer.gov/newscenter/newsfromnci/2011/ ReportNation2011Release.

99 *"Carson used DDT":* Pesticide Action Network, "The DDT Story." www.panna.org/issues/persistent-poisons/the-ddt-story.

99 *"wonder insecticide":* Edmund Russell, *War and Nature: Fighting Humans and Insects Using Chemicals from World War I to Silent Spring.* New York: Cambridge University Press, 2001.

100 *"to only a few chemicals":* Tina Rosenberg, NAS report cited in "What the World Needs Now Is DDT." *New York Times Magazine,*

April 11, 2004. www.acadiau.ca/~sskjei/cgi-bin/2713/Readings/rosenberg.pdf.

100 *"In 1820 about":* Francis Joseph Weiss, "Chemical Agriculture." *Scientific American,* August 1, 1952, 18.

101 *Agricultural productivity in the United States:* Fast Facts About Agriculture. American Farm Bureau Federation, 2014. www.fb.org/index.php?fuseaction=newsroom.fastfacts.

101 *first cases of evolving insect resistance:* Peter Jentsch, *Historical Perspectives on Fruit Production: Fruit Tree Pest Management, Regulation and New Chemistries.* Cornell University's Hudson Valley Lab. web.entomology.cornell.edu/jentsch/assets/historical-perspectives-on-apple-production.pdf.

102 *DDT is as carcinogenic:* International Agency for Research on Cancer (IARC), *IARC Monographs on the Evaluation of Carcinogenic Risks to Humans.* World Health Organization, March 31, 2014. monographs.iarc.fr/ENG/Classification/.

102 *largely dismissed the notion:* Deposition of Wilhelm C. Hueper, MD, May 24, 1957, *Lowe v. R.J. Reynolds Tobacco Co.* legacy.library.ucsf.edu/tid/vbk79a00;jsessionid=3F7F4CA89C0E4511C7A343999A512CC3.tobacco03.

104 *"that Americans born since":* Paul Ehrlich, "Eco-Catastrophe." *Ramparts,* September 1969, 26. www.unz.org/Pub/Ramparts-1969sep-00024.

104 *Fifty years later:* SEER Stat Fact Sheets: Leukemia. seer.cancer.gov/statfacts/html/leuks.html.

105 *relationship of cancer incidence to increasing age:* Robin P. Hertz, Margaret McDonald, and Kimary Kulig, *The Burden of Cancer in American Adults.* Pfizer Global Pharmaceuticals, 2003. www.pfizer.com/files/products/The_Burden_of_Cancer_in_American_Adults.pdf.

105 *The American Cancer Society reports:* "What Are the Key Statistics for Childhood Cancer?" American Cancer Society, January 31, 2014. www.cancer.org/cancer/cancerinchildren/detailedguide/cancer-in-children-key-statistics; Cancer Facts and Figures 2014, Special Section: "Cancer in Children and Adolescents." American Cancer Society,

2014. www.cancer.org/acs/groups/content/@research/documents/webcontent/acspc-041787.pdf.

106 *age-adjusted incidence rates:* Surveillance, Epidemiology, and End Results Program, SEER Cancer Statistics Review (1975–2009), Table 2.1, National Cancer Institute. seer.cancer.gov/archive/csr/1975_2009_pops09/browse_csr.php?sectionSEL=2&pageSEL=sect_02_Table.01.html.

106 *"is not believed to be causally related":* Eugenia Calle et al., "Organochlorines and Breast Cancer Risk." *CA: A Cancer Journal for Clinicians* 52.5 (September/October 2002): 301–309. onlinelibrary.wiley.com/doi/10.3322/canjclin.52.5.301/full.

106 *"Exposure to carcinogenic agents":* Cancer Facts and Figures 2014, American Cancer Society, 2014, 55. www.cancer.org/acs/groups/content/@research/documents/webcontent/acspc-042151.pdf.

107 *"Large organizations like":* "Harmful Substances—Chemicals, Pollution and Cancer," Cancer Research UK, accessed June 27, 2014. www.cancerresearchuk.org/cancer-info/healthyliving/harmfulsubstances/harmful-substances-chemicals-pollution-and-cancer.

107 *"so low that they":* Committee on Comparative Toxicity of Naturally Occurring Carcinogens, *Carcinogens and Anticarcinogens in the Human Diet: A Comparison of Naturally Occurring and Synthetic Substances.* Washington, DC: National Academies Press, 1996. www.nap.edu/openbook.php?record_id=5150&page=R1.

107 *cancers related to obesity:* Rudolf Kaaks and Tilman Kühn, "Epidemiology: Obesity and Cancer—The Evidence Is Fattening Up." *Nature Reviews Endocrinology* 10 (September 30, 2014), 644–645. www.nature.com/nrendo/journal/v10/n11/full/nrendo.2014.168.html; Nathan Berger, "Obesity and Cancer Pathogenesis." *Annals of the New York Academy of Sciences,* April 11, 2014, 57–76. onlinelibrary.wiley.com/doi/10.1111/nyas.12416/full; and Easter Vanni and Elisabetta Bugianesi, "Obesity and Liver Cancer." *Clinics in Liver Disease* 18.1 (February 2014), 191–203. www.sciencedirect.com/science/article/pii/S1089326113000627.

108 *"DDT is not":* J. Gordon Edwards, "DDT: A Case Study in Scientific

Fraud." *Journal of American Physicians and Surgeons,* Fall 2004, 86 [online]. www.jpands.org/vol9no3/edwards.pdf.

108 ***"The ultimate judgment":*** J. Gordon Edwards, "DDT: A Case Study in Scientific Fraud," 86.

108 ***"There is no objective reason":*** William Souder, *On a Farther Shore: The Life and Legacy of Rachel Carson, Author of Silent Spring.* New York: Broadway Books, 2013.

109 ***"tend to be morally suspicious":*** Dan M. Kahan et al., "The Polarizing Impact of Science Literacy and Numeracy on Perceived Climate Change Risks." *Nature Climate Change* 2 (August 16, 2012): 732–735. scholarship.law.gwu.edu/cgi/viewcontent.cgi?article=1298&context =faculty_publications.

110 ***"general lack of interest":*** Daniel W. Anderson, personal communication, 2004. reason.com/archives/2004/01/07/ddt-eggshells-and-me.

112 ***eggshell thinning of some bird species:*** Rhys E. Green, "Long-Term Decline in the Thickness of Eggshells of Thrushes, *Turdus* spp., in Britain." *Proceedings of the Royal Society B* 265 (April 22, 1998): 679–684. rspb.royalsocietypublishing.org/content/265/1397/679.short.

112 ***DDT did not cause eggshell:*** M. L. Scott et al., "Effects of PCBs, DDT, and Mercury Compounds Upon Egg Production, Hatchability and Shell Quality in Chickens and Japanese Quail." *Poultry Science* 54 (1975): 350–368. ps.oxfordjournals.org/content/54/2/350.short.

113 ***strict controls on DDT:*** Richard Tren, Richard Nchabi Kamwi, and Amir Attaran, "The UN Is Premature in Trying to Ban DDT for Malaria Control." *BMJ,* October 10, 2012, 345. www.bmj.com/content/ 345/bmj.e6801.

113 ***200 million people:*** Malaria Fact Sheet, World Health Organization, March 2014. www.who.int/mediacentre/factsheets/fs094/en/.

114 ***most searching inquiry ever:*** Renee Twombly, "Long Island Study Finds No Link Between Pollutants and Breast Cancer." *JNCI: Journal of the National Cancer Institute* 94.18 (September 17, 2002): 1348–1351. jnci.oxfordjournals.org/content/94/18/1348.full.

114 ***"are not associated":*** Gina Kolata and Deborah Winn cited in "Looking for the Link." *New York Times,* August 11, 2002. www.nytimes .com/2002/08/11/weekinreview/11KOLA.html.

115 ***longer a woman breast-fed:*** Torgil Möller et al., "Breast Cancer and

Breastfeeding : Collaborative Reanalysis of Individual Data from 47 Epidemiological Studies in 30 Countries, Including 50302 Women with Breast Cancer and 96973 Women Without the Disease." *Lancet* 360.9328 (July 2002): 187–195. lup.lub.lu.se/search/publication/1123899.

116 *"This only appears":* "Is There a Cancer 'Epidemic'?" Understanding Cancer Series, National Cancer Institute, January 28, 2005. www .cancer.gov/cancertopics/understandingcancer/cancer/page61.

116 *lifetime risk of:* American Cancer Society, "Lifetime Risk of Developing and Dying of Cancer," www.cancer.org/cancer/cancerbasics/ lifetime-probability-of-developing-or-dying-from-cancer.

116 *If you live long:* George Johnson, "Why Everyone Seems to Have Cancer." *New York Times,* January 4, 2014. www.nytimes.com/2014/ 01/05/sunday-review/why-everyone-seems-to-have-cancer.html?_r=0.

116 *annual death rates:* David S. Jones, Scott H. Podolsky, and Jeremy A. Greene, "The Burden of Disease and the Changing Task of Medicine." *New England Journal of Medicine* 366 (June 21, 2012): 2333–2338. www.nejm.org/doi/pdf/10.1056/NEJMp1113569.

116 *This initially sounds:* Felicitie C. *Bell and Michael L. Miller,* "Life Tables for the United States Social Security Area 1900–2100." Actuarial Study No. 120, Social Security Administration. www.ssa.gov/OACT/ NOTES/as120/LifeTables_Body.html.

116 *Today, 88 percent:* Elizabeth Arias, "United States Life Tables, 2009." National Vital Statistics Report, January 6, 2014. www.cdc.gov/nchs/ data/nvsr/nvsr62/nvsr62_07.pdf.

116 *In 1929, the first year:* Elizabeth Arias, "United States Life Tables, 2009."

117 *lung cancer death rate:* Morbidity and Mortality Weekly Report, "Achievements in Public Health, 1900–1999: Tobacco Use—United States, 1900–1999." Centers for Disease Control and Prevention, November 5, 1999, 986–993, www.cdc.gov/mmwr/preview/mmwrhtml/ mm4843a2.htm

117 *endocrine disrupter conjecture:* Howard A. Bern et al., "Statement from the Work Session on Chemically-Induced Alterations in Sexual Development: The Wildlife/Human Connection." Wingspread Conference Center, July 1991. www.ourstolenfuture.org/consensus/ wingspread1.htm.

118 *endocrine-related disorders:* Ake Bergman et al., *The State of the Science of Endocrine Disrupting Chemicals 2012: Summary for Decision-Makers.* United Nations Environment Programme and the World Health Organization, 2012, 2.

119 *Alarm about:* E. Carlsen et al., "Evidence for Decreasing Quality of Semen During Past 50 Years." *BMJ* 305 (1992): 609–613. www.ncbi .nlm.nih.gov/pmc/articles/PMC1883354/pdf/bmj00091-0019.pdf.

119 *falling sperm counts:* Steve Connor, "Why Be So Careless with the Facts?" *The Independent,* June 4, 1995. www.independent.co.uk/ voices/why-be-so-careless-with-the-facts-1585007.html.

119 *"allegations for a worldwide":* Harry Fisch and Stephen R. Braun, "Trends in Global Semen Parameter Values." *Asian Journal of Andrology* 15.2 (March 2013): 169–173. www.asiaandro.com/news/upload/ 20130912-aja2012143a.pdf.

119 *"generalized statements that hypospadias":* Suzan L. Carmichael, Gary M. Shaw, and Edward J. Lammer, "Environmental and Genetic Contributors to Hypospadias: A Review of the Epidemiologic Evidence." *Birth Defects Research Part A: Clinical and Molecular Teratology* 94.7 (July 2012): 499–510. www.ncbi.nlm.nih.gov/pmc/articles/ PMC3393839/pdf/nihms387688.pdf.

120 *"While genes involved":* L. F. M. van der Zanden et al., "Aetiology of Hypospadias: A Systematic Review of Genes and Environment." *Human Reproduction Update* 18.3 (February 26, 2012): 260–283.

120 *"A review of the epidemiologic data":* Harry Fisch, Grace Hyun, and Terry W. Hensle, "Rising Hypospadias Rates: Disproving a Myth." *The Journal of Pediatric Urology* 6 (2010): 37–39. www.pedclerk.sites .uchicago.edu/sites/pedclerk.uchicago.edu/files/uploads/1-s2.0 -S1477513109003490-main.pdf.

120 *"the trend of increasing adult height":* Fabrizio Giannandrea et al., "Case-Control Study of Anthropometric Measures and Testicular Cancer Risk." *Frontiers in Endocrinology* 3 (November 26, 2012): 144. www.ncbi.nlm.nih.gov/pmc/articles/PMC3505837/#!po=4.54545.

121 *"changes in diet":* Elizabeth E. Hatch et al., "Association of Endocrine Disruptors and Obesity: Perspectives from Epidemiologic Studies." *International Journal of Andrology* 33.2 (January 22, 2010): 1365–2605. www.ncbi.nlm.nih.gov/pmc/articles/PMC3005328/.

121 *changes in the number of calories:* Earl S. Ford and William H. Dietz, "Trends in Energy Intake Among Adults in the United States: Findings from NHANES." *American Journal of Clinical Nutrition* 97.4 (April 2013): 848–853.

121 *"Over the last 50 years":* T. S. Church et al., "Trends over 5 Decades in U.S. Occupation-Related Physical Activity and Their Associations with Obesity." *PLoS One* 6.5 (May 25, 2011): e19657. doi:10.1371/journal.pone.0019657.

121 *being overweight escalates:* Lei Chen, Dianna J. Magliano, and Paul Z. Zimmet, "The Worldwide Epidemiology of Type 2 Diabetes Mellitus— Present and Future Perspectives." *Nature Reviews Endocrinology* 8 (April 20, 2012): 228–236. 211.144.68.84:9998/91keshi/Public/File/ 34/-/pdf/nrendo.2011.183.pdf.

121 *"In no case was":* Kristina A. Thayer et al., "Role of Environmental Chemicals in Diabetes and Obesity: A National Toxicology Program Workshop Review," *Environmental Health Perspectives,* June 2012. ehp.niehs.nih.gov/wp-content/uploads/120/6/ehp.1104597.pdf.

121 *earlier onset of puberty:* Frank M. Biro et al., "Onset of Breast Development in a Longitudinal Cohort." *Pediatrics,* November 5, 2013. pediatrics.aappublications.org/content/early/2013/10/30/peds.2012 -3773; Paul B. Kaplowitz, "Link Between Body Fat and the Timing of Puberty." *Pediatrics,* February 1, 2008. pediatrics.aappublications.org/ content/121/Supplement_3/S208.long.

122 *"no evidence to suggest an increase":* Guilherme V. Polanczyk et al., "ADHD Prevalence Estimates Across Three Decades: An Updated Sys tematic Review and Meta-Regression Analysis." *International Journal of Epidemiology* 43.6 (January 24, 2014). ije.oxfordjournals.org/content/ 43/2/434.

122 *"hypothesis that the negligible exposure":* Gerhard J. Nohynek et al., "Endocrine Disruption: Fact or Urban Legend?" *Toxicology Letters* 223.3 (December 2013): 295–305.

123 *only 5 to 10 percent:* Gautam Naik and S. Stanley Young cited in "Analytical Trend Troubles Scientists." *Wall Street Journal,* May 4, 2012. online.wsj.com/news/articles/SB10001424052702303916904577377841 427001840; see also S. Stanley Young and Alan Karr, "Deming, Data and Observational Studies: A Process out of Control and Needing Fixing."

Significance, Journal of the Royal Statistical Society 8.3 (September 2011): 116–120. errorstatistics.files.wordpress.com/2014/04/young-karr-obs -study-problem.pdf.

123 *publish only studies with positive results:* Gary Taubes, "Epidemiology Faces Its Limits." *Science* 269.5221 (July 14, 1995), 169.

124 *"Investigators who find":* Taubes, "Epidemiology Faces Its Limits," 169.

124 *little thing from a big thing:* Taubes, "Epidemiology Faces Its Limits," 164.

124 *"only primitive tools":* Samuel Shapiro, "Looking to the 21st Century: Have We Learned from Our Mistakes, or Are We Doomed to Compound Them?" *Pharmacoepidemiology and Drug Safety* 13.4 (April 2004), 260.

124 *"pathological science":* Irving Langmuir and Robert N. Hall, "Pathological Science." *Physics Today,* October 1989. scitation.aip.org/ content/aip/magazine/physicstoday/article/42/10/10.1063/1.88120.5.

125 *"the first characteristic":* Denis Rousseau, "Case Studies in Pathological Science," *American Scientist* 80.1 (January–February 1992): 54–63.

125 *cell-based and animal model tests:* Laura N. Vandenberg et al., "Hormones and Endocrine-Disrupting Chemicals: Low-Dose Effects and Nonmonotonic Dose Responses." *Endocrine Reviews* 33.3 (June 2012): 378–455. www.ncbi.nlm.nih.gov/pmc/articles/PMC3365860/.

125 *not careful enough:* Frederick S. vom Saal et al., "Flawed Experimental Design Reveals the Need for Guidelines Requiring Appropriate Positive Controls in Endocrine Disruption Research." *Toxicological Sciences* 15.2 (February 17, 2010): 612–613. toxsci.oxfordjournals.org/content/ 115/2/612.short.

125 *impervious to critiques:* Leon E. Gray Jr. et al., "Rebuttal of 'Flawed Experimental Design Reveals the Need for Guidelines Requiring Appropriate Positive Controls in Endocrine Disruption Research' by vom Saal." *Toxicological Sciences* 115.2 (March 5, 2010): 614–620. toxsci .oxfordjournals.org/content/115/2/614.full#ref-68.

126 *"contradicts centuries of":* Nohynek et al., "Endocrine Disruption: Fact or Urban Legend?"

126 *"nutribogus epidemiology":* John P. A. Ioannidis, "Why Science Is

Not Necessarily Self-Correcting." *Perspectives on Psychological Science* 7.6 (November 2012): 645–654. 130.236.177.26/~729A94/mtrl/Why _science_is_not_necessarily_self-correcting.pdf.

126 ***"vested interest of scientists":*** Nohynek et al., "Endocrine Disruption: Fact or Urban Legend?"

5. The Attack of the Killer Tomatoes?

130 ***"overall cancer incidence rates":*** Report Prepared by the Hawai'i Tumor Registry for the Hawai'i State Department of Health: Kaua'i Cancer Cases, April 2013. health.hawaii.gov/wp-content/uploads/2013/09/Kauai-Cancer-Cases-April-2013.pdf.

130 ***"There's no real consensus":*** Doug Gurian-Sherman, "Are GMOs Worth the Trouble?" *MIT Technology Review,* March 27, 2014. www.technologyreview.com/view/525931/are-gmos-worth-the-trouble/.

131 ***a worldwide moratorium:*** Paul Berg et al., "Potential Biohazards of Recombinant DNA Molecules." Letter to the Editor, *Science* 185.4148 (July 16, 1974): 303.

131 ***"In the case of recombinant DNA":*** Liebe Cavalieri, "New Strains of Life—or Death." *New York Times Magazine,* August 22, 1976, 67.

132 ***"We want to be damned":*** Alfred Vellucci, cited in John Kifner, " 'Creation of Life' Experiment at Harvard Stirs Heated Debate." *New York Times,* June 17, 1976, 1.

132 ***"one of the world's major":*** Cambridge Community Development Department, "Cambridge: The Brains of Biotech, The Heart of Innovation," brochure. Downloaded 2014.

132 ***"Scientifically I was a nut":*** James Watson, personal communication to Ronald Bailey, cited in *Eco-Scam: The False Prophets of Ecological Apocalypse.* New York: St. Martin's Press, 1993, 94.

132 ***"In looking back":*** Burke Zimmerman, *Biofuture: Confronting the Genetic Era.* New York: Plenum Press, 1984, 176.

133 ***"The traditional notion":*** Ted Howard and Jeremy Rifkin, *Who Should Play God? The Artificial Creation of Life and What It Means for the Future of the Human Race.* New York: Delacorte Press, 1977, 223.

133 ***"Humanity seeks the elation":*** Jeremy Rifkin, *Algeny: A New Word—A New World.* New York: Penguin, 1984, 47.

133 *"as a cleverly constructed tract":* Stephen Jay Gould "Review of *Algeny."* *Discover,* January 1985, 34.

134 *"milk from treated cows":* FDA, Animal Veterinary, Product Safety Information, Bovine Somatotropin, January 28, 2014. www.fda.gov/AnimalVeterinary/SafetyHealth/ProductSafetyInformation/ucm055435.htm.

134 *regulators in Europe:* Ladina Caduff, "Growth Hormone and Beyond." Swiss Federal Institute of Technology Working Paper 8-2002.

135 *Bt protein is safe:* EPA, Bt Plant-Incorporated Protectants October 15, 2001: Biopesticides. Registration Action Document. www.epa.gov/oppbppd1/biopesticides/pips/bt_brad2/2-id_health.pdf.

135 *herbicide resistance trait:* EPA, Attachment III: Environmental Risk Assessment of Plant Incorporated Protectant (PIP) Inert Ingredients, December 2005. www.epa.gov/scipoly/sap/meetings/2005/december/pipinertenvironmentalriskassessment11-18-05.pdf.

135 *resource-poor farmers:* International Service for the Acquisition of Agri-biotech Applications, ISAAA Brief 46, "Global Status of Commercialized Biotech/GM Crops: 2013," Executive Summary, March 25, 2014. www.isaaa.org/resources/publications/briefs/46/executivesummary/default.asp.

136 *"led to strong distrust":* Sylvie Bonny, "Factors Explaining Opposition to GMOs in France and the Rest of Europe," in *Consumer Acceptance of Genetically Modified Foods,* Robert Evenson and Vittorio Santaniello, eds. Cambridge, MA: CABI Publishing, 2004, 181.

136 *"increased the public's attention":* Bonny, "Factors Explaining Opposition to GMOs in France and the Rest of Europe," 174.

137 *"symbolizes the negative aspects":* Bonny, "Factors Explaining Opposition to GMOs in France and the Rest of Europe," 183 184.

137 *"We call on the government":* Ranjit Devraj, "Health-India: Indian Cyclone Victims Guinea Pigs for U.S. Genetic Food." Interpress Service News Agency, June 12, 2000. www.ipsnews.net/2000/06/health-india-indian-cyclone-victims-guinea-pigs-for-usgenetic-food/.

137 *"To accuse the US":* Ronald Bailey, "Dr. Strangelunch." *Reason,* cited Per Pinstrup-Andersen, January 1, 2000. reason.com/archives/2001/01/01/dr-strangelunch.

138 *"We would rather starve":* Jennifer Cooke and Richard Downie, "African Perspectives on Genetically Modified Crops." A Report of the Center for Strategic and International Studies, Global Food Security Project, July 2010, 6. dspace.cigilibrary.org/jspui/bitstream/123456789/28948/1/African%20perspectives%20on%20genetically%20modified%20crops.pdf?1.

138 *"I would rather eat":* Davan Maharaj and Anthony Mukwita, "Zambia Rejects Gene-Altered U.S. Corn." *Los Angeles Times,* August 28, 2002. articles.latimes.com/2002/aug/28/world/fg-zambia28.

138 *Mwanawasa thought biotech crops:* Brooke Glass-O'Shea, "The History and Future of Genetically Modified Crops: Frankenfoods, Superweeds, and the Developing World." *Journal of Food Law and Policy* 7 (2011). papers.ssrn.com/sol3/papers.cfm?abstract_id=2019491.

138 *Brazilian member:* Ronald Bailey, "The Battle of Valle Verde." *Reason,* September 17, 2003. reason.com/archives/2003/09/17/the-battle-of-valle-verde.

139 *"not detected any significant hazard":* Alessandro Nicolia et al., "An Overview of the Last 10 Years of Genetically Engineered Crop Safety Research." *Critical Reviews in Biotechnology* 34.1 (March 2014): 77–88. www.geneticliteracyproject.org/wp/wp-content/uploads/2013/10/Nicolia-20131.pdf.

139 *"science is quite clear":* Board of Directors, Association for the Advancement of Science, "Statement by the AAAS Board of Directors on Labeling of Genetically Modified Foods," October 20, 2012.

139 *"no more risk in eating GMO food":* Anne Glover, cited in "No Risk with GMO Food, Says EU Chief Scientific Advisor." *Euractiv,* July 24, 2012. www.euractiv.com/innovation-enterprise/commission-science-supremo-endor-news-514072.

140 *"Bioengineered foods have been consumed":* Action of the AMA House of Delegates 2012 Annual Meeting: Council on Science and Public Health Report 2 Recommendations Adopted as Amended. hahaha.typepad.com/files/ama-on-bioengineered-foods.pdf.

140 *"no scientific evidence":* European Commission, "Commission Publishes Compendium of Results of EU-Funded Research on Genetically Modified Crops," December 9, 2010. europa.eu/rapid/press-release_IP-10

-1688_en.htm?locale=en; and European Commission, A Decade of EU-Funded GMO Research, December 2010. ec.europa.eu/research/biosociety/pdf/a_decade_of-eu-funded_gmo_research.pdf.

140 *"no adverse health effects":* National Research Council, Safety of Genetically Engineered Foods: Approaches to Assessing Unintended Health Effects. Washington, DC: National Academies Press, 2004 www.nap.edu/catalog.php?record_id=10977.

140 *"no evidence of any ill effects":* International Council for Science, "New Genetics, Food and Agriculture: Scientific Discoveries—Societal Dilemmas," 2003. www.icsu.org/publications/reports-and-reviews/new-genetics-food-and-agriculture-scientific-discoveries-societal-dilemas-2003/.

140 *"No effects on human health":* World Health Organization, "Frequently Asked Questions on Genetically Modified Foods," Food Safety. www.who.int/foodsafety/areas_work/food-technology/faq-genetically-modified-food/en/.

140 *"The level of safety":* Society of Toxicology, "The Safety of Genetically Modified Foods Produced Through Biotechnology," September 25, 2002. toxicology.org/gp/GM_Food.asp.

140 *"Biotechnology experts believe":* Government Accountability Office, *Genetically Modified Foods: Experts View Regimen of Safety Tests as Adequate, but FDA's Evaluation Process Could Be Enhanced,* May 23, 2002. www.gpo.gov/fdsys/pkg/GAOREPORTS-GAO-02-566/html/GAOREPORTS-GAO-02-566.htm.

141 *"no human health problems":* National Academy of Sciences, *Transgenic Plants and World Agriculture.* Washington, DC: National Academies Press, 2000. www.nap.edu/catalog.php?record_id=9889.

141 *"the technology itself":* Mark Bittman, "Leave 'Organic' Out of It." *New York Times,* May 6, 2014. www.nytimes.com/2014/05/07/opinion/bittman-leave-organic-out-of-it.html?hp&rref=opinion.

141 *mouse testicles blue:* Irina Ermakova, cited by Institute for Responsible Technology, "Genetically Modified Soy Linked to Sterility, Infant Mortality." www.responsibletechnology.org/article-gmo-soy-linked-to-sterility

141 *rats fed herbicide resistant:* Gilles-Éric Séralini et al.,"Long-Term Toxicity of a Roundup Herbicide and a Roundup-Tolerant Gene-

tically Modified Maize." *Food and Chemical Toxicology* 50.11 (November 2012): 4221–4231. www.sciencedirect.com/science/article/pii/S0278691512005637.

141 *"independent non-profit organization":* Committee for Independent Research and Information on Genetic Engineering, www.criigen.org/?option=com_content&task=blogcategory&id=52&Itemid=103.

141 *truly independent groups:* Frederic Schorsch, "Serious Inadequacies Regarding the Pathology Data Presented in the Paper by Séralini et al." *Food and Chemical Toxicology* 53 (March 2013): 465–466. www.sciencedirect.com/science/article/pii/S0278691512007880; and Erio Barale-Thomas, Letter to the Editor, *Food and Chemical Toxicology*, March 2013, 473–474. www.sciencedirect.com/science/article/pii/S0278691512007867.

142 *should never have published:* Andrew Revkin, DotEarth, *New York Times*, cited translation of French science academies' statement on the Seralini study. www.slideshare.net/Revkin/translation-of-french-science-academies-critique-of-controversial-gm-corn-study.

142 *"inadequately designed, analysed":* European Food Safety Authority, "Final Review of the Séralini et al. (2012a) Publication on a 2-Year Rodent Feeding Study with Glyphosate Formulations and GM Maize NK603 as Published Online on 19 September 2012 in Food and Chemical Toxicology." *EFSA Journal*, November 2012, 2986–2996. www.efsa.europa.eu/en/efsajournal/pub/2986.htm.

142 *decided to retract:* Retraction notice to "Long-Term Toxicity of a Roundup Herbicide and a Roundup-Tolerant Genetically Modified Maize." *Food and Chemical Toxicology*, January 2014. www.sciencedirect.com/science/article/pii/S0278691513008090.

142 *ban the importation of foods:* Emily Willingham, "Séralini Paper Influences Kenya Ban on GMO Imports." *Forbes*, December 9, 2012. www.forbes.com/sites/emilywillingham/2012/12/09/seralini-paper-influences-kenya-ban-of-gmo-imports/.

142 *decided to republish:* Gilles-Éric Séralini et al., "Republished Study: Long-Term Toxicity of a Roundup Herbicide and a Roundup-Tolerant Genetically Modified Maize." *Environmental Sciences Europe* 26.14 (June 24, 2014). www.enveurope.com/content/26/1/14.

143 *asserting that no such consensus exists:* European Network of

Scientists for Social and Environmental Responsibility, ENSSER Statement: "No Scientific Consensus on GMO Safety," October 23, 2013. www.ensser.org.

143 *fewer than three hundred scientists:* ENSSER List of Signatories as of December 2013. www.ensser.org/fileadmin/user_upload/signatories _as_of_131210_lv.pdf.

143 *"other synthetic herbicides":* Ralph E. Heimlich et al., "Genetically Engineered Crops: Has Adoption Reduced Pesticide Use?" *Agricultural Outlook,* August 2000, Economic Research Service, US Department of Agriculture, 13–17. www.agweb.com/assets/import/files/ao273f.pdf.

143 *very low toxicity:* Glyphosate Technical Fact Sheet, National Pesticide Information Center, npic.orst.edu/factsheets/glyphotech.pdf.

143 *breaks down quickly:* Adoption of Bioengineered Crops, Economic Research Service, US Department of Agriculture, AER-810, 26–29. www.ers.usda.gov/media/759233/aer810h_1_.pdf.

144 *"modestly increased":* Charles Benbrook, "Do GM Crops Mean Less Pesticide Use?" Mindfully.org, Pesticide Outlook, October 2001. www .mindfully.org/Pesticide/More-GMOs-Less-Pesticide.htm.

144 *biotech crops had reduced:* Ralph E. Heimlich et al., "Genetically Engineered Crops: Has Adoption Reduced Pesticide Use?" *Agricultural Outlook,* August 2000, Economic Research Service, US Department of Agriculture, 13–17. www.agweb.com/assets/import/files/ao273f.pdf.

144 *122 million pound increase:* Charles Benbrook, "Genetically Engineered Crops and Pesticide Use in the United States: The First Nine Years." BioTech InfoNet, Technical Paper Number 7, October 2004, 53. www.keine-gentechnik.de/bibliothek/anbau/studien/biotech _infonet_gvo_pestizide_041001.pdf.

144 *"reduced herbicide use":* Leonard P. Gianessi, "Economic and Herbicide Use Impacts of Glyphosate-Resistant Crops." *Pest Management* 61.3 (March 2005): 241–245. www.ask-force.org/web/Benefits/Gianessi -Benefits-2005.pdf.

144 *planting biotech crops:* Sujatha Sankula, "Quantification of the Impacts on US Agriculture of Biotechnology-Derived Crops Planted in 2005." National Center for Food and Agricultural Policy, November 2006, 110. www.ncfap.org/documents/2005biotechimpacts-finalversion .pdf.

144 *25 to 30 percent less herbicide:* Gijs Kleter et al., "Review: Altered Pesticide Use on Transgenic Crops and the Associated General Impact from an Environmental Perspective." *Pest Management Science* 63 (September 20, 2007): 1107–1115. www.cof.orst.edu/cof/teach/agbio2010/ Other%20Readings/Kleter%20Pestidice%20Use%20GM%20 Crop%20Rev%202007.pdf.

145 *increase of 383 million pounds:* Charles Benbrook, "Impacts of Genetically Engineered Crops on Pesticide Use in the United States: The First Thirteen Years." Organic Center Critical Issue Report, November 2009, 69. www.organic-center.org/reportfiles/GE13YearsReport.pdf.

145 *overall increase:* Charles Benbrook, "Impacts of Genetically Engineered Crops on Pesticide Use in the U.S.—The First Sixteen Years." *Environmental Sciences Europe* 24 (September 28, 2012). www .enveurope.com/content/24/1/24.

145 *reported these results:* Jon Entine, "Scientists Challenge Organic Backer Benbrook Claims That GM Crops Increase Pesticide Spraying," Genetic Literacy Project, October 12, 2012. www.geneticliteracy project.org/2012/10/12/scientists-journalists-challenge-organic -scientist-benbrook-claims-that-gm-crops increase-pesticide-spraying -harm-the-environment/.

145 *planting modern biotech crop varieties:* Graham Brookes and Peter Barfoot, "Global Impact of Biotech Crops: Environmental Effects, 1996– 2010." *GM Crops and Food: Biotechnology in Agriculture and the Food Chain,* April/May/June 2012, 129–137. www.landesbioscience.com/ journals/gmcrops/2012GMC0002R.pdf.

145 *national herbicide and insecticide usage:* Jorge Fernandez-Cornejo et al., "Pesticide Use in U.S. Agriculture: 21 Selected Crops, 1960–2008." Economic Information Bulletin No. (EIB-124), May 2014. www.ers .usda.gov/publications/eib-economic-information-bulletin/eib124 .aspx#.VD2XVRaKX_Y.

146 *147 agronomic studies:* Wilhelm Klümper and Matin Qaim, "A Meta-Analysis of the Impacts of Genetically Modified Crops." *PLoS One,* November 3, 2014. www.plosone.org/article/info%3Adoi%2F10.1371 %2Fjournal.pone.0111629.

147 *none reacted in a way:* Investigation of Human Health Effects Associated with Potential Exposure to Genetically Modified Corn, Centers

for Disease Control and Prevention, June 11, 2001, 24. www.cdc.gov/
nceh/ehhe/Cry9Creport/pdfs/cry9creport.pdf.

147 *lower in potent cancer-causing mycotoxins:* Felicia Wu, "Mycotoxin
Reduction in Bt Corn: Potential Economic, Health, and Regulatory
Impacts." *ISB News Report,* September 2006, 3. www.nbiap.vt.edu/
news/2006/artspdf/sep0604.pdf.

148 *poisoned monarch butterfly caterpillars:* John H. Losey, Linda S.
Rayor, and Maureen E. Carter, "Transgenic Pollen Harms Monarch
Larvae." *Nature* 399 (May 20, 1999): 214. www.nature.com/nature/
journal/v399/n6733/abs/399214a0.html.

148 *impact on monarch butterfly populations:* Mark K. Searset et al.,
"Impact of Bt Corn Pollen on Monarch Butterfly Populations: A Risk
Assessment." *Proceedings of the National Academy of Sciences* 98.21
(2001): 11937–11942. www.pnas.org/content/98/21/11937.long.

148 *"commercialized GM crops":* J. E. Carpenter, "Impact of GM Crops on
Biodiversity." *GM Crops* 2.1 (January/March 2011): 7–23. www.ncbi
.nlm.nih.gov/pubmed/21844695.

148 *"Many U.S. farmers":* National Research Council, *The Impact of
Genetically Engineered Crops on Farm Sustainability in the United
States.* Washington, DC: National Academies Press, 2010. www
.nationalacademies.org/onpinews/newsitem.aspx?RecordID=12804.

149 *"no evidence":* International Council for Science, "New Genetics, Food
and Agriculture: Scientific Discoveries—Societal Dilemmas," 2003.
www.icsu.org/publications/reports-and-reviews/new-genetics-food
-and-agriculture-scientific-discoveries-societal-dilemas-2003.

149 *introduction of modern herbicides and pesticides:* Ronald Bailey,
"Asking the Wrong Questions." *Reason,* November 5, 2003. reason
.com/archives/2003/11/05/asking-the-wrong-questions. Cites results
of three-year Farm Scale Evaluation reported in Theme Issue "The
Farm Scale Evaluations of Spring-Sown Genetically Modified Crops" of
Philosophical Transactions of the Royal Society B, November 29, 2003.
rstb.royalsocietypublishing.org/content/358/1439.toc.

150 *saves 1 billion tons of topsoil:* Richard Fawcett and Dan Towery,
*Conservation Tillage and Plant Biotechnology: How New Technologies
Can Improve the Environment by Reducing the Need to Plow.* Conser-

vation Technology Information Center, 2003. www.ctic.org/media/
pdf/Biotech2003.pdf.

150 *reduces the runoff:* Fawcett and Towery, *Conservation Tillage and
Plant Biotechnology.*

150 *"marked contrast to yield increases":* "An Analysis of 'Failure to Yield'
by Doug Gurian-Sherman, Union of Concerned Scientists," Wayne
Parrott, Institute of Plant Breeding, Genetics and Genomics, and Department of Crop and Soil Sciences, University of Georgia. Updated April 2,
2010. www.salmone.org/wp-content/uploads/2013/06/response-to-ucs
.pdf.

151 *"are scale neutral":* FAO, *The State of Food and Agriculture 2003–2004.*
www.fao.org/es/esa/pdf/sofa_flyer_04_en.pdf.

151 *insect-resistant cotton varieties:* Richard Bennett et al., "Farm-Level
Economic Performance of Genetically Modified Cotton in Maharash-
tra, India." *Review of Agricultural Economics* 28.1 (January 2006): 59–71.
www.agbioworld.org/pdf/ReviewAgricEconomicsj.pdf.

151 *"adopters compared to non-adopters":* Janet E. Carpenter, "Peer-
Reviewed Surveys Indicate Positive Impact of Commercialized GM
Crops." *Nature Biotechnology* 28.4 (April 2010): 319–321. www.ask-force
.org/web/Benefits/Carpenter-Peer-Reviewed-Surveys-GM-crops-2010
.pdf.

151 *"non-target and beneficial organisms":* A. M. Mannion and Stephen
Morse, "Gm Crops 1996–2012: A Review of Agronomic, Environmental and Socio-Economic Impacts," University of Surrey, Centre for
Environmental Strategy Working Paper 04/13; also published as University of Reading Geographical Paper No. 195, April 2013. www
.surrey.ac.uk/ces/activity/publications/index.htm.

152 *"traditional pillars of sustainability":* Julian Park et al., "The Role
of Transgenic Crops in Sustainable Development." *Plant Biotechnology Journal* 9.1 (January 2011): 2–21.

152 *"positive in both developed and developing":* A. M. Mannion and
Stephen Morse, "Biotechnology in Agriculture: Agronomic and Environmental Considerations and Reflections Based on 15 Years of GM
Crops." *Progress in Physical Geography* 36.6 (December 2012): 747–
763. ppg.sagepub.com/content/36/6/747.abstract.

153 *anti-biotech activists:* Mark Lynas, "The True Story About Who Destroyed a Genetically Modified Rice Crop." *Slate,* August 26, 2013. www.slate.com/blogs/Future_tense/2013/08/26/golden_rice _attack_in_philippines_anti_gmo_activists_lie_about_protest _and.html.

153 *attacks on crop biotechnology:* Bruce Alberts et al., "Standing Up for GMOs." *Science* 341.6152 (September 20, 2013): 1320. www.science mag.org/content/341/6152/1320.full

153 *1.4 million life-years:* Justus Wesseler and David Zilberman, "The Economic Power of the Golden Rice Opposition." *Environment and Development Economics* 19.6 (December 2014): 724–742. journals .cambridge.org/action/displayAbstract;jsessionid=1BDDA1658A3F79A 7C0EECD575E3C90CD.journals?aid=9402215&fileId=S1355770X 1300065X.

154 *"The first inkling of what":* Andrew J. Forgash, "History, Evolution, and Consequences of Insecticide Resistance." *Pesticide Biochemistry and Physiology* 22.2 (October 1984): 178–186. www.sciencedirect.com/ science/article/pii/0048357584900877.

155 *"A mixture of insecticides":* Mallet, James, "The Evolution of Insecticide Resistance: Have the Insects Won?" *Trends in Ecology and Evolution* 4.11 (November 1989): 336–340. www.ucl.ac.uk/taxome/jim/ pap/mallet89tree.pdf.

155 *herbicide-resistant weeds:* International Survey of Herbicide Resistant Weeds, April 2014, www.weedscience.org/summary/home.aspx.

156 *"Herbicide resistant weed development":* Andrew Kniss, "Where Are the Superweeds?" May 1, 2013. weedcontrolfreaks.com/2013/05/ superweed/.

156 *organic standards are process standards:* Ronald Bailey, "Organic Law." *Reason,* October 2, 2002. reason.com/archives/2002/10/02/organic -law. Cites Kershen's legal analysis.

158 *"The gradual spread of sterility":* Vandana Shiva, *Stolen Harvest: The Hijacking of the Global Food Supply.* Brooklyn, NY: South End Press, 2000.

158 *"the fuel of* interest": Abraham Lincoln, "Lecture on Discoveries and Inventions," April 6, 1858. www.abrahamlincolnonline.org/lincoln/ speeches/discoveries.htm.

159 *court decided against Schmeiser:* Monsanto Canada Inc. v. *Schmeiser* [2004], 1 S.C.R. 902, 2004 SCC 34. scc-csc.lexum.com/scc-csc/scc-csc/en/item/2147/index.do.

160 *court ruled unanimously against:* Bowman v. Monsanto, US Supreme Court, May 13, 2013. www.supremecourt.gov/opinions/12pdf/11-796_c07d.pdf.

161 *"most people do not actively avoid":* European Commission, *A Decade of EU-Funded GMO Research,* December 2010. ec.europa.eu/research/biosociety/pdf/a_decade_of_eu funded_gmo_research.pdf.

161 *"has no basis for concluding":* FDA, "Guidance for Industry: Voluntary Labeling Indicating Whether Foods Have or Have Not Been Developed Using Bioengineering; Draft Guidance," January 2001. www.fda.gov/Food/GuidanceRegulation/GuidanceDocumentsRegulatoryInformation/LabelingNutrition/ucm059098.htm.

161 *"The FDA does not require labeling":* Board of Directors, Association for the Advancement of Science, "Statement by the AAAS Board of Directors on Labeling of Genetically Modified Foods," October 20, 2012.

162 *no scientific reason:* Editorial Board, "Why Label Genetically Engineered Food?" *New York Times,* March 14, 2013. www.nytimes.com/2013/03/15/opinion/why-label-genetically-engineered-food.html?_r=1&.

162 *source non-GMO ingredients:* Walter Robb and A. C. Gallo, "GMO Labeling Coming to Whole Foods Market." Whole Story Blog, March 8, 2013. www.wholefoodsmarket.com/blog/gmo-labeling-coming-whole-foods-market.

162 *require the FDA to promulgate rules:* Safe and Accurate Food Labeling Act of 2014, pompeo.house.gov/uploadedfiles/safeandaccuratefoodlabellingactof2014.pdf.

162 *"slow the development of agricultural biotechnology":* David Zilberman, "Why Labeling of GMOs Is Actually Bad for People and the Environment." blogs.berkeley.edu/2012/06/06/why-labeling-of-gmos-is-actually-bad-for-people-and-the-environment/comment-page-2/.

163 *costs about $136 million:* Phillips McDougall, "The Cost and Time Involved in the Discovery, Development and Authorisation of a New Plant Biotechnology Derived Trait." A Consultancy Study for Crop Life International, September 2011.

164 *"Indian farmers have committed suicide":* Amy Goodman, Interview with Vandana Shiva at *Democracy Now!* Transcript, "Vandana Shiva on International Women's Day: 'Capitalist Patriarchy Has Aggravated Violence Against Women,'" March 8, 2013. www.democracynow.org/2013/3/8/vandana_shiva_on_intl_womens_day.

164 *Kloor eviscerates Shiva's stories:* Keith Kloor, "The GMO-Suicide Myth." *Issues in Science and Technology,* February 5, 2014. issues.org/30-2/keith/.

165 *"the adoption of GM cotton":* Matin Qaim and Shahzad Kouser, "Genetically Modified Crops and Food Security." *PLoS One,* June 15, 2013, doi:10.1371/journal.pone.0064879.

165 *"did not find a systematic relation":* Anoop Sadanandan, "Political Economy of Suicide: Financial Reforms, Credit Crunches, and Farmer Suicides in India." *Journal of Developing Areas* 48.4 (Fall 2014): 287–307. www.anoopsadanandan.com/PoliticalEconomyofSuicide.pdf.

165 *"male farmer suicide rates have actually declined":* Ian Plewis, "Hard Evidence: Does GM Cotton Lead to Farmer Suicide in India?" *The Conversation,* March 12, 2014. theconversation.com/hard-evidence-does-gm-cotton-lead-to-farmer-suicide-in-india-24045.

166 *"specific changes in wood chemistry":* Steven Strauss, Stephen P. DiFazio, and Richard Meilan, "Genetically Modified Poplars in Context." *The Forestry Chronicle* 77.2 (March/April 2001): 271–276. www.cof.orst.edu/coops/tbgrc/publications/Strauss_2001_The_Forestry_Chronicle.pdf.

166 *most of the world's wood products:* Roger Sedjo, "From Foraging to Cropping: The Transition to Plantation Forestry, and Implications for Wood Supply and Demand." Resources for the Future, www.fao.org/docrep/003/x8820e/x8820e06.htm.

167 *"wheat-gene tweaked freaks":* Bernd Heinrich, "Revitalizing Our Forests." *New York Times,* December 21, 2013.

167 *new biotech variety of rice:* Elizabeth Svoboda, "The Future of Farming Is Nitrogen Efficiency." *Fast Company,* October 1, 2008. www.fastcompany.com/1007058/future-farming-nitrogen-efficiency.

167 *biotech canola:* Arcadia Biosciences website, April 20, 2014. www.arcadiabio.com/products.

167 *salt tolerance, drought tolerance:* Andy Coghlan, "Super-rice Defies

Triple Whammy of Stresses." *The New Scientist* 16 (February 28, 2014): 38. www.newscientist.com/article/dn25147-superrice-defies-triple -whammy-of-stresses.html#.U1KcvFfJGRM.

168 *transfer the C4 photosynthetic pathway:* Shanta Karki, Govinda Rizal, and William Paul Quick, "Improvement of Photosynthesis in Rice (Oryza sativa L.) by Inserting the C4 Pathway." *Rice Journal,* October 28, 2013. www.thericejournal.com/content/6/1/28/.

168 *heat tolerance:* Elizabeth A. Ainsworth, Alistair Rogers and Andrew D. B. Leakey, "Targets for Crop Biotechnology in a Future High-CO_2 and High-O_3 World." Plant Physiology *147.1* (May 2008): 13–19. www.plantphysiol.org/content/147/1/13.full.

168 *viral disease resistance:* Donald Danforth Plant Science Center, "Danforth Center Licenses Technology from Dow Agro Science to Improve Important Staple Crop," Cassava Mosaic Disease, July 15, 2011. www .danforthcenter.org/news-media/news-releases-10/Danforth-Center -Licenses-Technology-from-Dow-Agro-Science-to-Improve-Important -Staple-Crop.

168 *fungal disease resistance:* Beat Keller, Wilhelm Gruissem, Michael Winzeler, Franz Bigler, and Fabio Mascher, Switzerland—Resistance against Fungal Diseases in Wheat—University and ETH Zurich, Agroscope Research Stations ART and ACW, Greenbiotech Briefing, greenbiotech.eu/briefing-paper/public-research-stopped-or-moved -abroad/switzerland-resistance-against-fungal-diseases-in-wheat -university-and-eth-zurich-agroscope-research-stations-art-and-acw/.

168 *bacterial disease resistance:* Jonathan D. G. Jones et al., "Elevating Crop Disease Resistance with Cloned Genes." *Philosophical Transactions of the Royal Society B,* April 5, 2014. rstb.royalsocietypublishing .org/content/369/1639/20130087.full.html.

168 *algae that can suck carbon dioxide:* D. Ryan Georgianna and Stephen P. Mayfield, "Exploiting Diversity and Synthetic Biology for the Production of Algal Biofuels." *Nature* 488 (August 16, 2012): 329–335. labs.biology.ucsd.edu/schroeder/bggn227/2014%20Lectures/Mayfield/ Algae%20biofuels%20review.pdf.

169 *enviropigs:* Enviropig, University of Guelph, www.uoguelph.ca/ enviropig/.

169 *AquAdvantage Salmon:* FDA, Center for Veterinary Medicine, United

States Food and Drug Administration, Department of Health and Human Services, "Preliminary Finding of No Significant Impact (FONSI) for AquAdvantage Salmon," May 4, 2012. www.fda.gov/downloads/AnimalVeterinary/DevelopmentApprovalProcess/GeneticEngineering/GeneticallyEngineeredAnimals/UCM333105.pdf.

169 *environmental organizations oppose this technology:* Friends of the Earth News release, " 'Fatally Flawed' FDA Assessment to Unleash Genetically Engineered Salmon," December 21, 2012. www.foe.org/news/news-releases/2012-12-fatally-flawed-fda-assessment-to-unleash-genetically-engineered-salmon.

169 *introduced legislation to ban the biotech salmon:* Paul Voosen, "House Moves to Ban Modified Salmon." *New York Times,* June 16, 2011. www.nytimes.com/gwire/2011/06/16/16greenwire-house-moves-to-ban-modified-salmon-84165.html.

169 *hopes to bring its animal-free milk:* Linda Qiu, "Milk Grown in a Lab Is Humane and Sustainable. But Will It Catch On?" *National Geographic,* October 22, 2014. news.nationalgeographic.com/news/2014/10/141022-lab-grown-milk-biotechnology-gmo-food-climate/.

169 *dairy cattle emit 17 percent:* Dario Caro et al., "Global and Regional Trends in Greenhouse Gas Emissions from Livestock." Climatic Change 126 (2014): 203–216. DOI: 10.1007/s10584-014-1197-x.

170 *organic agriculture as currently practiced:* H. L. Tuomisto et al., "Does Organic Farming Reduce Environmental Impacts? A Meta-Analysis of European Research." *Journal of Environmental Management* 112 (September 1, 2012): 309–320. www.fraw.org.uk/files/food/tuomisto_2012.pdf; Verena Seufert, Navin Ramankutty, and Jonathan A. Foley, "Comparing the Yields of Organic and Conventional Agriculture," *Nature* 485 (May 10, 2012): 229–234. serenoregis.org/wp-content/uploads/2012/06/nature11069.pdf.

170 *judicious incorporation of two important strands:* Pamela Ronald and Raoul Adamchak, *Tomorrow's Table: Organic Farming, Genetics, and the Future of Food.* New York: Oxford University Press, 2008, xi.

6. Can We Cope with the Heat?

172 *"a true planetary emergency"*: MSNBC, "Gore Takes Warming Warning to Congress," March 21, 2007. www.nbcnews.com/id/17718399/ns/us_news-environment/t/gore-takes-warming-warning-congress/#.VEBxrBaKX_Y.

172 *"an endless chain of disasters"*: Bill McKibben, *Oil and Honey: The Education of an Unlikely Activist*. New York: Times Books, 2013.

172 *"Our economic system"*: Naomi Klein, *This Changes Everything: Capitalism vs. the Climate*. New York: Simon & Schuster, 2014.

172 *"the key new text"*: Bill McKibben, "An Actual Exit from Climate Hell." *The Dish*, August 28, 2014. dish.andrewsullivan.com/author/billmckibbendish/.

173 *"man-made global warming"*: John Gizzi, John, "Inhofe Was First to Declare Global Warming 'The Greatest Hoax.'" *Human Events*, August 6, 2012. humanevents.com/2012/08/06/inhofe-was-first-to-declare-global-warming-the-greatest-hoax/.

173 *"The scientific reality is"*: Marc Morano, "Submitted Written Testimony of Marc Morano, publisher of Climate Depot and former staff of US Senate Environment and Public Works Committee," Congressional Field Hearing: "The Origins and Response to Climate Change," May 30, 2013. www.climatedepot.com/2013/05/31/submitted-written testimony-of-climate-depots-marc-morano-at-congressional-hearing-on-climate-change-the-origins-and-response-to-climate-change/.

173 *"There is no convincing evidence"*: Fred Singer in Larry Bell, "Any Global Warming Since 1978? Two Climate Experts Debate This." *Forbes*, June 18, 2013. www.forbes.com/sites/larrybell/2013/06/18/any-global-warming-since-1978-two-climate-experts-debate-this/.

174 *chance that 2014 was warmer:* National Oceanic and Atmospheric Administration, National Climatic Data Center, "Global Summary Information—December 2014," January 2015, www.ncdc.noaa.gov/sotc/summary-info/global/2014/12.

175 *"Warming of the climate system"*: IPCC, *Climate Change 2013: The Physical Science Basis*, 2013, www.ipcc.ch/report/ar5/wg1/; Summary for Policymakers, www.climatechange2013.org/images/report/WG1AR5

_SPM_FINAL.pdf; Technical Summary, www.climatechange2013
.org/images/report/WG1AR5_TS_FINAL.pdf.

175 *restated and bolstered:* IPCC, *Climate Change 2014: Synthesis Report,*
November 2014, www.ipcc.ch/pdf/assessment-report/ar5/syr/SYR
_AR5_LONGERREPORT.pdf.

175 *vast majority of climate researchers:* Stacy Rosenberg et al., *Climate
Change: A Profile of U.S. Climate Scientists' Perspectives.* Institute for
Science, Technology, and Public Policy, Texas A&M University, 2009.
bush.tamu.edu/istpp/news/story/8/ClimateScientistsPerspectives
ClimaticChange.pdf; and John Cook et al., "Quantifying the Consensus
on Anthropogenic Global Warming in the Scientific Literature." *Environmental Research Letters* 8.2 (May 15, 2013). iopscience.iop.org/1748
-9326/8/2/024024.

176 *most mountain glaciers:* I. Velicogna, T. C. Sutterley, and M. R. van
den Broeke, "Regional Acceleration in Ice Mass Loss from Greenland
and Antarctica Using GRACE Time-Variable Gravity Data." *Geophysical Research Letters* 41.22 (November 28, 2014): 8130–8137.

177 *sea level could at worst:* S. Jevrejeva, A. Grinsted, and J. C. Moore,
"Upper Limit for Sea Level Projections by 2100." *Environmental
Research Letters* 9.10 (October 10, 2014). iopscience.iop.org/1748-9326/9/
10/104008/.

177 *global average sea level:* Kurt Lambeck et al., "Sea Level and Global
Ice Volumes from the Last Glacial Maximum to the Holocene." *Proceedings of the National Academy of Sciences* 111.43 (October 13, 2014).
www.pnas.org/content/early/2014/10/08/1411762111.abstract.

178 *40% to 70% reduction in GHG:* IPCC, *Climate Change 2014: Synthesis
Report,* November 2014, www.ipcc.ch/pdf/assessment-report/ar5/syr/
SYR_AR5_LONGERREPORT.pdf.

178 *"The science is settled":* Andrea Seabrooke, "Gore Takes Global Warming Message to Congress." NPR, March 21, 2007. www.npr.org/
templates/story/story.php?storyId=9047642.

178 *"human activity is responsible":* Robin Bravender, "EPA Chief Goes
Toe-to-Toe with Senate GOP over Climate Science." *New York Times,* February 23, 2010. www.nytimes.com/gwire/2010/02/23/23greenwire-epa
-chief-goes-toe-to-toe-with-senate-gop-over-72892.html.

178 *global temperature has been essentially flat:* Ross McKitrick, "HAC-

Robust Measurement of the Duration of a Trendless Subsample in a Global Climate Time Series." *Open Journal of Statistics* 4 (August 2014): 527–535. www.scirp.org/journal/PaperInformation.aspx?paperID=49307.

179 *"The observed rate of warming":* John C. Fyfe, Nathan P. Gillett, and Francis W. Ziers, "Overestimated Gobal Warming over the Past 20 Years." *Nature Climate Change* 3 (August 28, 2013): 767–769. www .nature.com/nclimate/journal/v3/n9/full/nclimate1972.html?WT.ec _id=NCLIMATE-201309.

180 *"that is where models show":* John Christy, personal communication, 2013.

180 *The private research group:* Remote Sensing Systems, *Climate Analysis,* accessed February 11, 2015, www.remss.com/research/ climate.

181 *decreases in stratospheric water vapor:* Susan Solomon et al., "Contributions of Stratospheric Water Vapor to Decadal Changes in the Rate of Global Warming." *Science* 327.5970 (January 28, 2010): 1219–1223. www.sciencemag.org/content/327/5970/1219.

181 *the missing heat is supposedly hiding:* Gerald A. Meehl. et al., "Model-Based Evidence of Deep-Ocean Heat Uptake During Surface-Temperature Hiatus Periods." *Nature Climate Change* 1 (September 18, 2011): 360–364. echorock.cgd.ucar.edu/staff/trenbert/trenberth .papers/Meehl_Natureclimatechange2011-1.pdf.

181 *a prolonged solar minimum:* James Hansen et al., "Earth's Energy Imbalance and Implications." *Atmospheric Chemistry and Physics* 11 (2011): 13421–13449. www.atmos-chem-phys.net/11/13421/2011/acp-11 -13421-2011.html.

181 *Pacific Ocean trade winds:* Matthew H. England et al., "Recent Intensification of Wind-Driven Circulation in the Pacific and the Ongoing Warming Hiatus." *Nature Climate Change* 4 (February 9, 2014): 222–227. www.nature.com/nclimate/journal/v4/n3/abs/ncli mate2106.html.

181 *changes in North Atlantic Ocean circulation:* Xianyao Chen and Ka-Kit Tung, "Varying Planetary Heat Sink Led to Global-Warming Slowdown and Acceleration." *Science* 345.6199 (August 22, 2014): 897–903. www.sisal.unam.mx/labeco/LAB_ECOLOGIA/OF_files/ heat%20sink%20led%20to%20global-warming%20slowdown.pdf.

181 *natural variations in Pacific trade winds:* Masahiro Watanabe et al., "Contribution of Natural Decadal Variability to Global Warming Acceleration and Hiatus." *Nature Climate Change* 4 (August 31, 2014): 893–897. www.nature.com/nclimate/journal/v4/n10/abs/nclimate 2355.html.

182 *little chance of a hiatus decade:* Nicola Maher, Alexander Sen Gupta, and Matthew England, "Drivers of Decadal Hiatus Periods in the 20th and 21st Centuries." *Geophysical Research Letters* 41.16 (August 28, 2014): 5978–5986. onlinelibrary.wiley.com/enhanced/doi/10.1002/ 2014GL060527/#Survey.

182 *volcanic particles on global atmospheric temperatures:* D. A. Ridley et al., "Total Volcanic Stratospheric Aerosol Optical Depths and Implications for Global Climate Change." *Geophysical Research Letters* 41.22 (October 31, 2014): 7763–7769. onlinelibrary.wiley.com/doi/ 10.1002/2014GL061541/abstract;jsessionid=E1B1AB1E8B14434278D7 CFC307E735A4.f01t01.

182 *sulfuric acid particles from small volcanic eruptions:* Benjamin D. Santer et al., "Observed Multi-Variable Signals of Late 20th and Early 21st Century Volcanic Activity." *Geophysical Research Letters,* January 2014; DOI: 10.1002/2014GL062366.

182 *"We do not find that aerosols":* A. Gettleman, D. T. Shindell, and J. F. Lamarque, "Impact of Aerosol Radiative Effects on 2000–2010 Surface Temperatures." *Climate Dynamics,* January 2015; www.cgd.ucar.edu/ staff/andrew/papers/gettelman2015aerorecent.pdf.

182 *man-made aerosols might be responsible:* Gavin Schmidt, Drew Shindell, and Kostas Tsigaridas, "Reconciling Warming Trends." *Nature Geoscience* 7 (March 2014): 158–160. www.blc.arizona.edu/ courses/schaffer/182h/Climate/Reconciling%20Warming%20Trends .pdf.

182 *ocean abyss below 2,000 meters:* William Llovel et al., "Deep-Ocean Contribution to Sea Level and Energy Budget Not Detectable over the Past Decade." *Nature Climate Change* 4 (October 5, 2014): 1031–1035. www.nature.com/nclimate/journal/vaop/ncurrent/full/nclimate 2387.html.

183 *"The cold waters of Earth's deep ocean":* Jet Propulsion Laboratory, "NASA Study Finds Earth's Ocean Abyss Has Not Warmed." Califor-

nia Institute of Technology, October 6, 2014, www.jpl.nasa.gov/news/news.php?feature=4321.

183 *upper layers of the southern oceans:* Paul J. Duracket et al., "Quantifying Underestimates of Long-Term Upper-Ocean Warming." *Nature Climate Change* 4 (October 5, 2014): 999–1005. www.nature.com/nclimate/journal/vaop/ncurrent/full/nclimate2389.html.

183 *observed global mean surface air temperature:* H. Douville, A. Voldoire, and O. Geoffroy, "The Recent Global-Warming Hiatus: What Is the Role of Pacific Variability?" *Geophysical Research Letters,* January 2015, onlinelibrary.wiley.com/doi/10.1002/2014GL062775/abstract.

183 *thirty-four of the climate models:* Patrick T. Brown, Wenhong Li, and Shang-Ping Xie, "Regions of Significant Influence on Unforced Global Mean Surface Air Temperature Variability in Climate Models." *Journal of Geophysical Research: Atmospheres,* January 2015, online library.wiley.com/doi/10.1002/2014JD022576/abstract.

183 *why global mean surface temperatures:* Tim Lucas, "Climate Models Disagree On Why Temperature Wiggles Occur." Duke University environment press release, January 26, 2015, nicholas.duke.edu/news/climate-models-disagree-why-temperature-wiggles-occur.

184 *results of comparing the outputs:* Jochem Marotzke and Piers M. Forster, "Forcing, Feedback and Internal Variability in Global Temperature Trends." *Nature* 517 (January 29, 2015) 565–570. www.nature.com/nature/journal/v517/n7536/full/nature14117.html?WT.ec_id=NATURE-20150129#author-information.

184 *"they don't get them at the right time":* Piers Forster cited by Reporting ClimateScience.com, "Study: Models Not to Blame in Failure to Predict Pause," January 28, 2015, www.reportingclimatescience.com/news-stories/article/models-not-to-blame-in-failure-to-predict-pause-says-study.html.

184 *What natural fluctuations:* Bryon A. Steinman, Michael E. Mann, Sonya K. Miller, "Atlantic and Pacific multidecadal oscillations and Northern Hemisphere temperatures," Science, February 27, 2015, 988–991, www.sciencemag.org/content/347/6225/988.

185 *"ocean warming dominates":* IPCC, *Climate Change 2014: Synthesis Report,* November 2014, 6. www.ipcc.ch/pdf/assessment-report/ar5/syr/SYR_AR5_LONGERREPORT.pdf.

185 *natural internal variability:* Judith Curry, "The IPCC's Inconvenient Truth." *Climate, Etc.,* September 20, 2013. judithcurry.com/2013/09/20/the-ipccs-inconvenient-truth/.

186 **Researchers from the Pacific Northwest:** Steven J. Smith, James Edmonds, Corinne A. Hartin, Anupriya Mundra and Katherine Calvin, "Near-term acceleration in the rate of temperature change," *Nature Climate Change,* March 9, 2015, www.nature.com/nclimate/journal/vaop/ncurrent/full/nclimate2552.html.

186 *mild El Niño—like conditions:* Climate Prediction Center/NCEP, "ESNO: Recent Evolution, Current Status and Predictions," January 19, 2015, www.cpc.ncep.noaa.gov/products/Analysis_monitoring/lanina/enso_evolution-status-fcsts-web.pdf.

188 **"most likely value of equilibrium climate sensitivity":** A. Otto et al., "Energy Budget Constraints on Climate Response." *Nature Geoscience* 6.6 (June 2013): 415–416. eprints.whiterose.ac.uk/76064/7/ngeo1836%281%29_with_coversheet.pdf.

188 *"is 1.8°C, with 90% C.I.":* R. B. Skeie et al., "A Lower and More Constrained Estimate of Climate Sensitivity Using Updated Observations and Detailed Radiative Forcing Time Series." *Earth Systems Dynamics* 5 (March 25, 2014): 139–175. www.earth-syst-dynam.net/5/139/2014/esd-5-139-2014.html.

189 *calculated a transient climate response:* Drew Shindell, "Inhomogeneous Forcing and Transient Climate Sensitivity." *Nature Climate Change* 4 (March 9, 2014): 274–277. www.nature.com/nclimate/journal/v4/n4/nclimate2136/metrics.

188 *researchers at Texas A&M University:* J. R. Kummer and A. E. Dessler, "The Impact of Forcing Efficacy on the Equilibrium Climate Sensitivity." *Geophysical Research Letters* 41.10 (May 28, 2014): 3565–3568. onlinelibrary.wiley.com/doi/10.1002/2014GL060046/abstract.

189 *the best estimate for climate sensitivity:* Nicholas Lewis and Judith Curry, "The Implications for Climate Sensitivity of AR5 Forcing and Heat Uptake Estimates." *Climate Dynamics,* September 2014, 1–15. link.springer.com/article/10.1007/s00382-014-2342-y.

189 *twenty years of temperature observations:* Nathan M. Urban et al., "Historical and Future Learning About Climate Sensitivity." *Geo-*

physical *Research Letters* 41.7 (April 16, 2014): 2543–2552. onlinelibrary
.wiley.com/doi/10.1002/2014GL059484/abstract.

190 ***"Impacts of ocean acidification":*** IPCC, *Climate Change 2014:*
Impacts, Adaptation, and Vulnerability, April 2014, Chapter 6, 138.
ipcc-wg2.gov/AR5/images/uploads/WGIIAR5-Chap6_FGDall.pdf.

190 ***as acidity increases:*** K. L. Ricke et al., "Risks to Coral Reefs from
Ocean Carbon Chemistry Changes in Recent Earth Systems Model Projections." *Environmental Research Letters,* July 3, 2013, 6. iopscience
.iop.org/1748-9326/8/3/034003/pdf/1748-9326_8_3_034003.pdf.

190 ***corals might reach a tipping point:*** O. Hoegh-Guldberg et al., "Coral
Reefs Under Rapid Climate Change and Ocean Acidification." *Science*
318.5857 (December 2007): 1737–1742. www.geneseo.edu/~bosch/Hoegh
-Guldberg.pdf.

190 ***tropical reefs might not be affected:*** S. Comeau et al., "The Responses
of Eight Coral Reef Calcifiers to Increasing Partial Pressure of CO_2 Do
Not Exhibit a Tipping Point." *Limnology and Oceanography* 58.1 (January 2013): 388–398. www.aslo.info/lo/toc/vol_58/issue_1/0388.pdf.

191 ***cold-water Mediterranean corals:*** C. Maier et al., "Respiration of
Mediterranean Cold-Water Corals Is Not Affected by Ocean Acidification
as Projected for the End of the Century." *Biogeosciences* 10 (August 27,
2013): 5671–5680, biogeosciences.net/10/5671/2013/bg-10-5671-2013
.pdf; see also S. J. Hennige et al., "Short-Term Metabolic and Growth
Responses of the Cold-Water Coral Lophelia pertusa to Ocean Acidification." *Deep Sea Research Part II: Topical Studies in Oceanography*
99 (January 2014): 27–35. www.sciencedirect.com/science/article/pii/
S0967064513002774.

191 ***overall effects on marine organisms:*** Astrid C. Wittman and Hans-O.
Pörtner, "Sensitivities of Extant Animal Taxa to Ocean Acidification."
Nature Climate Change 3 (August 25, 2013): 995–1001, www.nature
.com/nclimate/journal/v3/n11/full/nclimate1982.html; and also,
Kristy J. Kroeker et al., "Impacts of Ocean Acidification on Marine
Organisms: Quantifying Sensitivities and Interaction with Warming."
Global Change Biology 19.6 (June 2013): 1884–1896. www.ncbi.nlm
.nih.gov/pmc/articles/PMC3664023/.

191 ***damages in 2095 are $12 trillion:*** William D. Nordhaus, "Economic

Aspects of Global Warming in a Post-Copenhagen Environment." *Proceedings of the National Academy of Sciences* 107.26 (May 10, 2010): 11721–11726. www.pnas.org/content/107/26/11721.full.pdf+html&.

192 *how the world's economy might evolve:* International Institute for Applied Systems Analysis, SSP Database Version 0.9.3, revised March 2013, secure.iiasa.ac.at/web-apps/ene/SspDb/dsd?Action=html page&page=about.

193 *an average of about 1.5 percent:* William Nordhaus, *The Climate Casino: Risk, Uncertainty, and Economics for a Warming World.* New Haven: Yale University Press, 2013, 139.

194 *business-as-usual path:* Nicholas Stern, *Stern Review: The Economics of Climate Change,* Executive Summary, 2006. siteresources.world bank.org/INTINDONESIA/Resources/226271-1170911056314/3428109-1174614780539/SternReviewEng.pdf.

194 *"rich generations have a lower ethical claim":* Nordhaus, *The Climate Casino,* 187.

195 *extreme weather:* Michael Bastasch, "Boxer Uses Oklahoma Tornado to Push Carbon Tax." *The Daily Caller,* May 21, 2013. dailycaller.com/2013/05/21/boxer-uses-okla-tornado-to-push-carbon-tax/.

195 *destruction caused by Superstorm Sandy:* Greenpeace, "Hurricane Sandy = Climate Change," Extreme Weather and Climate Change, 2013. www.greenpeace.org/usa/en/campaigns/global-warming-and-energy/Extreme-Weather-and-Climate-Change/.

196 *hurricanes, typhoons, hailstorms, or tornadoes:* IPCC, *Managing the Risks of Extreme Events and Disasters to Advance Climate Change Adaptation. A Special Report of Working Groups I and II of the Intergovernmental Panel on Climate Change.* New York: Cambridge University Press, 2012. www.ipcc-wg2.gov/SREX/.

196 *economic losses from weather- and climate-related disasters:* IPCC, *Climate Change 2014: Synthesis Report,* November 2014, 16. www.ipcc.ch/pdf/assessment-report/ar5/syr/SYR_AR5_LONGERREPORT.pdf.

196 *"there has been little change in drought":* Justin Sheffield, Eric F. Wood, and Michael Roderick, "Little Change in Global Drought over the Past 60 Years." *Nature* 491 (November 14, 2012): 435–438. www

.nature.com/nature/journal/v491/n7424/full/nature11575.html?WT
.ec_id=NATURE-20121115.

196 *not a factor in the extreme drought:* Richard Seager et al., "Causes and Predictability of the 2011–14 California Drought." NOAA Drought Task Force, December 2014. cpo.noaa.gov/sites/cpo/MAPP/Task%20Forces/DTF/californiadrought/california_drought_report.pdf.

197 *"has declined by more than 90 percent":* Indur Goklany, Wealth and Safety: The Amazing Decline in Deaths from Extreme Weather in an Era of Global Warming, 1900–2010. Reason Foundation, September 2011. reason.org/files/deaths_from_extreme_weather_1900_2010.pdf.

198 *"anthropogenic climate change so far":* Laurens Bouwer, "Have Disaster Losses Increased Due to Anthropogenic Climate Change?" *Bulletin of the American Meteorological Society,* January 27, 2011, 39–46. journals.ametsoc.org/doi/pdf/10.1175/2010BAMS3092.1.

199 *"same result for all disasters":* Eric Neumayer and Fabian Barthel, "Normalizing Economic Loss from Natural Disasters: A Global Analysis (December 5, 2010)." *Global Environmental Change* 21.1 (2011) 13–24. Available at SSRN: ssrn.com/abstract=1720414.

199 *"Results show no detectable sign":* J. I. Barredo, "Normalised Flood Losses in Europe 1970–2006." *Natural Hazards and Earth Systems Sciences* 9 (February 9, 2009): 97–104, www.nat-hazards-earth-syst-sci.net/9/97/2009/nhess-9-97-2009.pdf; and J. I. Barredo, "No Upward Trend in Normalised Windstorm Losses in Europe: 1970–2008," *Natural Hazards and Earth Systems Sciences* 10 (January 15, 2010): 97–104. www.nat-hazards-earth-syst-sci.net/9/97/2009/nhess-9-97-2009.pdf.

201 *global greenhouse gas emissions:* IPCC, *Climate Change 2014: Mitigation of Climate Change,* www.ipcc.ch/report/ar5/wg3/; Summary for Policymakers, report.mitigation2014.org/spm/ipcc_wg3_ar5_summary-for-policymakers_approved.pdf; Technical Summary, report.mitigation2014.org/drafts/final-draft-postplenary/ipcc_wg3_ar5_final-draft_postplenary_technical-summary.pdf.

202 *each country made pledges:* US-China Joint Announcement on Climate Change, White House, November 11, 2014. www.whitehouse.gov/the-press-office/2014/11/11/us-china-joint-announcement-climate-change.

205 *preliminary draft document:* UNFCCC, *The Lima Call for Climate Action*, Decision-/CP.20, unfccc.int/files/meetings/lima_dec_2014/application/pdf/auv_cop20_lima_call_for_climate_action.pdf.

206 *permit prices had risen:* Ewa Krukowska, "EON Urges EU Policy Revamp as Power Market Faces Crisis," Bloomberg, May 6, 2014, www.bloomberg.com/news/2014-05-06/eon-urges-eu-policy-revamp-as-power-markets-face-crisis.html.

206 *Far under the price:* Stanley Reed, "European Lawmakers Try to Spur Market for Carbon Emissions Credits." *New York Times*, February 6, 2014. www.nytimes.com/2014/02/07/business/international/european-lawmakers-try-to-spur-market-for-carbon-emission-credits.html?_r=0.

208 *will cost European consumers:* Michael Szabo and Jeff Coelho, "EUAs Could Crash to 3 Euros Next Year, Says UBS." *Climate Justice Now*, November 18, 2011, www.climate-justice-now.org/euas-could-crash-to-3-euros-next-year-says-ubs/.

209 *"natural baseline is a zero-carbon-tax level of emissions":* William D. Nordhaus, "After Kyoto: Alternative Mechanisms to Control Global Warming." *Foreign Policy in Focus*, March 26, 2006. fpif.org/after_kyoto_alternative_mechanisms_to_control_global_warming/.

211 *government consumption subsidies for fossil fuels:* International Energy Agency, Energy Subsidies, *World Energy Outlook*. www.worldenergyoutlook.org/resources/energysubsidies/.

211 *barmy to subsidize agriculture:* "Farmgate: The Developmental Impact of Agricultural Subsidies," ActionAid, 2012. www.actionaid.org.uk/sites/default/files/content_document/farmgate_3132004_12159.pdf.

213 *"Prudence demands that we consider":* Ken Caldeira, "We Should Plan for the Worst-Case Climate Scenario." *The Bulletin of the Atomic Scientists*, July 29, 2008, thebulletin.org/has-time-come-geoengineering/we-should-plan-worst-case-climate-scenario.

214 *The National Academy of Sciences:* Committee on Geoengineering Climate: National Research Council, *Climate Intervention: Reflecting Sunlight to Cool the Earth*, February 2015, National Academy of Sciences Press, 234 pp. www.nap.edu/catalog/18988/climate-intervention-reflecting-sunlight-to-cool-earth; and Committee on Geoengineering Climate: National Research Council, *Climate Intervention: Carbon*

Dioxide Removal and Reliable Sequestration, National Academy of Science Press, February 2015, 140 pp. www.nap.edu./catalog/18805/climate-intervention-carbon-dioxide-removal-and-reliable-sequestration.

215 **bioenergy carbon capture and storage:** Elmar Kriegler et al., "Is Atmospheric Carbon Dioxide Removal a Game Changer for Climate Change Mitigation?" *Climatic Change* 118.1 (May 2013): 45–57. link .springer.com/article/10.1007%2Fs10584-012-0681-4.

215 **Another proposal is direct air capture:** Klaus Lackner et al., "The Urgency of the Development of CO_2 Capture from Ambient Air." *Proceedings of the National Academy of Sciences* 109.33 (June 28, 2012): 13156–13162. www.pnas.org/content/early/2012/07/26/1108765109; and Robert Kunzig, "Scrubbing the Skies," *National Geographic,* ngm .nationalgeographic.com/big-idea/13/carbon-capture-pg2

215 **"such research is a dangerous distraction":** Hands Off Mother Earth. Letter in opposition to the Stratospheric Particle Injection for Climate Engineering (SPICE) project. Sent to Chris Huhne, Secretary of State for Energy and Climate Change, September 26, 2011. www.handsoff motherearth.org/wp-content/uploads/2011/09/SPICE-Opposition -Letter.pdf.

215 **"If humans perceive an easy technological fix":** Alan Robock, "20 Reasons Why Geoengineering May Be a Bad Idea." *Bulletin of the Atomic Scientists* 64.2 (May/June 2008): 14–18. climate.envsci.rutgers .edu/pdf/20Reasons.pdf

215 **geoengineering would likely shift rainfall patterns:** Scott Barrett et al., "Climate Engineering Reconsidered." *Nature Climate Change* 4 (June 25, 2014): 527–529. www.nature.com/nclimate/journal/v4/n7/ full/nclimate2278.html; and also Daniela F. Cusack et al., "An Interdisciplinary Assessment of Climate Engineering Strategies." *Frontiers in Ecology and the Environment* 12.5 (June 2014): 280–287. www .esajournals.org/doi/abs/10.1890/130030.

216 **cost-benefit analysis should not apply:** Martin Weitzman, "On Modeling and Interpreting the Economics of Catastrophic Climate Change." *Review of Economics and Statistics* 91.1 (February 2009): 1–19. www .mitpressjournals.org/doi/abs/10.1162/rest.91.1.1#.U7HYa6goxyg.

217 **a persuasive critique of Weitzman's dismal conclusions:** William Nordhaus, "An Analysis of the Dismal Theorem," Cowles Foundation

Discussion Paper No. 1686, January 16, 2009. cowles.econ.yale.edu/P/
cd/d16b/d1686.pdf; and William Nordhaus, "Economic Policy in the
Face of Severe Tail Events." *Journal of Public Economic Theory*, 14.2
(2012): 197–219, www.econ.yale.edu/~nordhaus/homepage/documents/
Nordhaus_TailEvents_JPET_2012.pdf.

219   *the more scientifically literate:* Dan M. Kahan et al., "The Tragedy
      of the Risk-Perception Commons: Culture Conflict, Rationality Con-
      flict, and Climate Change" (2011). Temple University Legal Studies
      Research Paper No. 2011-26; Cultural Cognition Project Working Paper
      No. 89; Yale Law and Economics Research Paper No. 435; Yale Law
      School, Public Law Working Paper No. 230. Available at SSRN: ssrn
      .com/abstract=1871503 or dx.doi.org/10.2139/ssrn.1871503.

222   *stop the development:* James Gustave Speth, *Red Sky at Morning: Ame-*
      rica and the Global Environmental Crisis. New Haven, CT: Yale University
      Press, 2004; and Ronald Bailey, "The Cultural Contradictions of Envi-
      ronmentalism: Fast Breeder Reactor Edition." *Reason*, October 7, 2009.

223   *coal generation kills about 4,000 times:* Jerome Roos, "Coal Kills
      4,000 Times More People Per Unit of Energy Than Does Nuclear."
      Breakthrough Institute, April 11, 2011. thebreakthrough.org/archive/
      coal_kills_4000_times_more_peo.

223   *nuclear power avoided:* Pushker A. Kharecha and James E. Hansen,
      "Prevented Mortality and Greenhouse Gas Emissions from Historical
      and Projected Nuclear Power." *Environmental Science and Technology* 47
      (March 15, 2013): 4889–4895. pubs.acs.org/doi/full/10.1021/es3051197.

224   *"would decay to background levels":* Albert J. Juhasz, Richard A.
      Rarick, and Rajmohan Rangarajan, "High Efficiency Nuclear Power
      Plants Using Liquid Fluoride Thorium Reactor Technology." Seventh
      International Energy Conversion Engineering Conference, American
      Institute of Aeronautics and Astronautics, August 2–5, 2009, Denver,
      Colorado, enu.kz/repository/2009/AIAA-2009-4565.pdf.

224   *China is working on a project:* Jennifer Duggan, "China Working on
      Uranium-Free Nuclear Plants in Attempt to Combat Smog." *The Guard-*
      ian, March 19, 2014. www.theguardian.com/world/2014/mar/19/
      china-uranium-nuclear-plants-smog-thorium.

224   *traveling wave reactors are designed:* Tyler Ellis, "Traveling-Wave
      Reactors: A Truly Sustainable and Full-Scale Resource for Global

Energy Needs," Paper 10189, Proceedings of ICAPP 2010, San Diego, CA, USA, June 13–17, 2010. large.stanford.edu/courses/2012/ph241/levin2/docs/ICAPP_2010_Paper_10189.pdf.

225 *supply enough electricity to run a small city:* Guy Norris, "Skunk Works Reveals Compact Nuclear Fusion Reactor Details," *Aviation Week & Space Technology,* October 15, 2014. aviationweek.com/technology/skunk-works-reveals-compact-fusion-reactor-details.

226 *solar, geothermal, and wind energy:* Energy Information Administration, *Monthly Energy Review: December 2014,* Electricity Net Generation: Total (All Sectors), Table 7.2a, www.eia.gov/totalenergy/data/monthly/pdf/mer.pdf.

226 *"large-scale moon colonization":* Megan Nicholson and Matthew Stepp, "Challenging the Clean Energy Deployment Consensus." Information Technology and Innovation Foundation, October 23, 2013. www2.itif.org/2013-challenging-clean-energy-deployment-consensus.pdf.

227 *In a 2011 paper, the Stanford engineer Mark Jacobson:* Mark Z. Jacobson and Mark A Delucchi, "Providing All Global Energy with Wind, Water, and Solar Power, Part 1: Technologies, Energy Resources, Quantities and Areas of Infrastructure, and Materials." *Energy Policy* 39 (2011): 1154–1169; see especially 1160. old.rgo.ru/wp-content/uploads/2011/12/JDEnPolicyPt1.pdf.

228 *some way to store electricity: Annual Energy Outlook 2014,* Energy Information Administration, "Levelized Cost and Levelized Avoided Cost of New Generation Resources in the Annual Energy Outlook 2014," April 17, 2014, www.eia.gov/forecasts/aeo/electricity_generation.cfm.

228 *two to three times more generating capacity:* Cory Budischak et al., "Cost-Minimized Combinations of Wind Power, Solar Power, and Electrochemical Storage, Powering the Grid Up to 99.9% of the Time." *Journal of Power Sources* 225 (March 1, 2013): 60–74. www.ceoe.udel.edu/windpower/resources/BudischakEtAl-AsPublished-Corrected.pdf.

229 *Americans were willing to pay just under $10 per month:* Ed Crooks, "Voters Put $10 Limit on Green Energy Cost." *The Financial Times,* June 17, 2011. www.financialexpress.com/news/Voters-put-10-limit-on-green-energy-cost/804824.

229 *"Despite the skepticism of experts":* Vivek Wadhwa, "The Coming Era of Unlimited—and Free—Clean Energy." *Washington Post,*

September 19, 2014. www.washingtonpost.com/blogs/innovations/
wp/2014/09/19/the-coming-era-of-unlimited-and-free-clean-energy/.

230 *levelized unsubsidized cost of utility-scale solar PV:* Lazard, *Lazard's
Levelized Cost of Energy Analysis—Version 8.0,* September 2014, www
.lazard.com/PDF/Levelized%20Cost%20of%20Energy%20-%20
Version%208.0.pdf.

230 *"still require conventional technologies":* George Bilicic cited in press
release, "Lazard Releases New Levelized Cost of Energy Analysis," September 18, 2014. www.marketwatch.com/story/lazard-releases-new
-levelized-cost-of-energy-analysis-2014-09-18.

231 *low-end levelized cost for solar PV:* Electric Power Research Institute,
Integrated Generation Technology Options 2012, February 19, 2013.
www.epri.com/abstracts/Pages/ProductAbstract.aspx?productId
=000000000001026656.

231 *solar PV will be $101 per megawatt-hour:* Energy Information
Administration, *Annual Energy Outlook 2014,* May 7, 2014. www.eia
.gov/forecasts/aeo/electricity_generation.cfm.

231 *figure is already 15.9 gigawatts:* Solar Energy Industries Association, "Over Half a Million Solar Installations Now Online in the U.S.,"
Solar Energy Facts: Q2 2014. www.seia.org/sites/default/files/Q2%20
2014%20SMI%20Fact%20Sheet_0.pdf.

231 *global production capacity of solar cells/modules:* Selya Price and
Robert Margolis, "2008 Solar Technologies Market Report," National
Renewal Energy Laboratory, US Department of Energy, January 2010, 17.
www.nrel.gov/tech_deployment/pdfs/2008_solar_market_report.pdf.

231 *85 gigawatts in 2016:* Mike Munsell, "Polysilicon Capacity Growth to
Accelerate, Enabling 85GW of Solar Panel Production in 2016." Green-
TechMedia, October 14, 2014. www.greentechmedia.com/articles/read/
Polysilicon-Capacity-Growth-to-Accelerate-Enabling-85-GW-of-Solar
-Panel-Pr.

232 *That would not be too cheap:* Lewis L. Strauss, speech at National
Association of Science Writers, September 16, 1954, www.thisday
inquotes.com/2009/09/too-cheap-to-meter-nuclear-quote-debate
.html.

232 *disruptive new innovations:* Seth Fletcher, "Secretive Company
Claims Battery Breakthrough." *Scientific American,* August 20, 2014.

www.scientificamerican.com/article/secretive-company-claims
-battery-breakthrough/.

232 **There will be no further global treaties:** Ronald Bailey, "The Kyoto
Protocol Is Dead." *Reason,* December 17, 2004. reason.com/archives/
2004/12/17/the-kyoto-protocol-is-dead.

233 **"the international community should stop chasing the chimera":**
Timothy Wirth and Thomas Daschle, "A Blueprint to End Paralysis
Over Global Action on Climate." Y*ale Environment 360,* May 19, 2014.
e360.yale.edu/feature/a_blueprint_to_end_paralysis_over_global
_action_on_climate/2766/.

234 **"when policies focused on economic growth":** Roger Pielke Jr., *The
Climate Fix: What Scientists and Politicians Won't Tell You About
Global Warming. New York:* Basic Books, 2010, 272.

234 **"The paramount goal of climate policy":** Matthew Stepp and Megan
Nicholson, *Beyond 2015: An Innovation-Based Framework for Global Climate Policy.* Center for Clean Energy Innovation, Information Technology and Innovation Foundation, May 2014.

234 **"Social and environmental hazards like climate change":** Mark
Caine et al., *Our High Energy Planet—A Climate Pragmatism Pro-
ject.* Breakthrough Institute, April 2014; and Charles R. Frank Jr., *The
Net Benefits of Low and No-Carbon Electricity Technologies.* Working
Paper 73, Brookings Institution, May 2014.

234 **"Societies that are able to meet their energy needs":** Mark Caine et al.,
Our High Energy Planet: A Climate Pragmatism Project. April 2014.
thebreakthrough.org/images/pdfs/Our-High-Energy-Planet.pdf.

236 **a system of electricity regulation:** R. Richard Geddes, "A Historical
Perspective on Electric Utility Regulation." *Regulation,* Winter 1992.

236 **have amounted to more than $837 billion:** *60 Years of Energy Incen-
tives: An Analysis of Federal Expenditures for Energy Development,* Management Information Services, October 2011.

7. Is the Ark Sinking?

240 **"A large fraction of both":** Summary for Policymakers, *Climate Change
2014: Impacts, Adaptation, and Vulnerability,* IPCC March 31, 2014, 15.

240 **"Current rates of extinction":** S. L. Pimm et al. "The Biodiversity of

Species and Their Rates of Extinction, Distribution, and Protection."
Science 344.6187 (May 30, 2014). www.sciencemag.org/content/344/
6187/1246752.

240 *"It could be a scary future":* Center for Biological Diversity, "The
Extinction Crisis," accessed May 30, 2014. www.biologicaldiversity.org/
programs/biodiversity/elements_of_biodiversity/extinction_crisis/.

240 *"We're destroying the rest of life":* E. O. Wilson, April 30, 2012. "E. O.
Wilson wants to know why you're not protesting in the streets." Interview with Lisa Hymas, *Grist.*

240 *"We're destroying the rest of life":* Barnosky cited in "World's Sixth
Mass Extinction May Be Underway—Study." *The Independent*, March
7, 2011, www.independent.co.uk/environment/worlds-sixth-mass
-extinction-may-be-underway—study-2234388.html; see also Anthony
Barnosky et al., "Has the Sixth Mass Extinction Already Arrived?"
Nature 471 (March 3, 2011): 51–57. ib.berkeley.edu/labs/barnosky/
Barnosky%20et%20al%20Sixth%20Extinction%20Nature.pdf.

241 *somewhere between 75 and 80 percent:* Gaylord Nelson, quoting
Dr. S. Dillon Ripley, *Look*, April 1970.

242 *709 known species as having gone extinct:* IUCN Red List of Threatened Species, accessed May 30, 2014. www.iucnredlist.org/.

242 *322 species have become extinct:* Rodolfo Dirzo et al., "Defaunation
in the Anthropocene." *Science* 345.6195 (July 25, 2014): 401–406. www
.sciencemag.org/content/345/6195/401.

242 *half the number of vertebrates:* Richard McLellan, ed., *Living Planet
Report 2014*. World Wildlife Fund, 2014. wwf.panda.org/about_our
_earth/all_publications/living_planet_report/.

243 *108 species as having gone extinct:* Kieran Suckling, Rhiwena Slack,
and Brian Nowicki, "Extinction and the Endangered Species Act."
Center for Biological Diversity, May 1, 2004. www.biologicaldiversity
.org/publications/papers/ExtinctAndESA.pdf.

243 *giant Palouse earthworm:* Doug Zimmer, "Giant Palouse Earthworm
Not Warranted for ESA Protections." Pacific Region News Release, US
Fish and Wildlife Service, July 25, 2011, www.fws.gov/pacific/news/
news.cfm?id=2144374846.

243 *Some 178 mammal species:* Anthony Barnosky, PNAS, "Megafauna
Biomass Tradeoff as a Driver of Quaternary and Future Extinctions."

Proceedings of the National Academy of Science 105, supp. 1 (August 12, 2008). www.pnas.org/content/105/suppl.1/11543.full.

244 *after humans arrived in North America:* K. J. Willis and G. M. MacDonald, "Long-Term Ecological Records and Their Relevance to Climate Change Predictions for a Warmer World." *Annual Review of Ecology, Evolution, and Systematics* 42 (August 23, 2011): 267–287. www.annualreviews.org/doi/abs/10.1146/annurev-ecolsys-102209 -144704.

244 *Polynesian wayfarers caused the extinction:* Jeremy Hance, "Humans Killed over 10 Percent of the World's Bird Species When They Colonized the Pacific Islands." *Mongabay,* March 25, 2013. news.mongabay.com/ 2013/0325-hance-bird-extinction-pacific.html.

245 *"extinctions caused by habitat loss":* Fangliang He and Stephen P. Hubbell, "Species-Area Relationships Always Overestimate Extinction Rates from Habitat Loss." *Nature* 473 (May 19, 2011): 368–371. www .nature.com/nature/journal/v473/n7347/full/nature09985.html.

245 *"Models project that the risk":* IPCC, *Climate Change 2014: Impacts, Adaptation and Vulnerability,* Chapter 4: "Terrestrial and Inland Water Systems," March 31, 2014, 153. ipcc-wg2.gov/AR5/images/uploads/ WGIIAR5-Chap4_FGDall.pdf.

246 *protected areas have nearly doubled:* World Bank, *World Development Indicators: Deforestation and Biodiversity,* accessed June 3, 2014, wdi .worldbank.org/table/3.4.

246 *what is happening to marine biodiversity:* Douglas J. McCauley et al., "Marine Defaunation: Animal Loss in the Global Ocean." *Science* 347.6219 (January 16, 2015). www.sciencepubs.org/content/347/ 6219/1255641.abstract.

247 *populations of large open ocean predators:* Jeremy B. C. Jackson, "Ecological Extinction and Evolution in the Brave New Ocean," *Proceedings of the National Academy of Sciences* 105, supp. 1 (August 12, 2008): 11458– 11465. www.pnas.org/content/105/Supplement_1/11458.full.

247 *"halts, and even reverses":* Christopher Costello, Steven D. Gaines, and John Lynham, "Can Catch Shares Prevent Fisheries Collapse?" *Science* 321.5896 (September 19, 2008): 1678–1681. www.sciencemag.org/ content/321/5896/1678.short .

248 *"The city is the most environmentally benign form":* Stewart Brand,

citing Peter Calthorpe in "How Slums Can Save the Planet." *Prospect,* January 27, 2010. www.prospectmagazine.co.uk/magazine/how-slums -can-save-the-planet/#.U7sUp6goxyg.

249 *a globally interconnected world:* Paolo D'Odorico et al., "Feeding Humanity Through Global Food Trade." *Earth's Future* 2.9 (September 2014): 458–469. onlinelibrary.wiley.com/doi/10.1002/2014EF000250/ abstract.

249 *"we will need to find a way to reintegrate":* Jeremy Rifkin, "The Risks of Too Much City." *Washington Post,* December 17, 2006. www .washingtonpost.com/wp-dyn/content/article/2006/12/15/ AR2006121501647.html.

249 *"We believe that projecting conservative values":* Jesse H. Ausubel, Iddo K. Wernick, and Paul E. Waggoner, "Peak Farmland and the Prospect for Land Sparing." *Population Development Review* 38, supplement 1 (February 2013): 221–242. onlinelibrary.wiley.com/doi/10.1111/j.1728 -4457.2013.00561.x/pdf.

250 *"among 50 nations with extensive forests":* Pekka E. Kauppi et al., "Returning Forests Analyzed with the Forest Identity." *Proceedings of the National Academy of Sciences* 103.46 (November 13, 2006): 17574– 17579. www.pnas.org/content/103/46/17574.short.

250 *In 2014:* Remi D'Annunzio, Erik J. Lindquist, and Kenneth G. MacDicken, "Global forest land-use change from 1990 to 2010: an update to a global remote sensing survey of forests," FAO, 2014, www.lafranceagricole.fr/ var/gfa/storage/fichiers-pdf/Docs/2014/forest-FAO.pdf.

250 *A February 2015 study:* Do-Hyung Kim, Joseph O. Sexton, John R. Townshend, "Accelerated Deforestation in the Humid Tropics from 1990s to the 2000s," Geophysical Research Letters, February 11, 2015, onlinelibrary.wiley.com/enhanced/doi/10.1002/2014GL062777/ ?campaign=wlytk-41855.5282060185.

251 *secondary forests in tropical Africa, America, and Asia:* Food and Agriculture Organization, *The State of the World's Forests 2005.* United Nations, 2005. www.fao.org/docrep/007/y5574e/y5574e00.htm.

251 *"doubtful that more than 10% of the tropical forests":* R. Dirzo and P. H. Raven, "Global State of Biodiversity and Loss." *Annual Review of Environment and Resources* 28 (November 2003): 137–167.

251 *current forest trends suggest:* Joseph Wright, "The Future of Tropi-

cal Forests," *Annals of the New York Academy of Sciences,* Vol. 1195: *The Year in Ecology and Conservation Biology,* 2010, 27.

252 ***"The increase in secondary forest":*** Eldredge Bermingham, "Will Tropical Species Survive?" *Biological Conservation Newsletter,* Smithsonian Natural History Museum, February 2009. botany.si.edu/pubs/bcn/issue/pdf/bcn290.pdf.

252 ***"Surprisingly, few species":*** Louise Gray, citing Nigel Stork in "Extinction of Millions of Species 'Greatly Exaggerated.'" *Telegraph,* January 23, 2013. www.telegraph.co.uk/earth/earthnews/9824723/Extinction-of-millions-of-species-greatly-exaggerated.html; see also Mark J. Costello, Robert M. May, and Nigel E. Stork, "Can We Name Earth's Species Before They Go Extinct?" *Science* 339.6118 (January 25, 2013): 413–416. bug.tamu.edu/entocourses/ento601/pdf/Costello_et_al_2013.pdf.

252 ***farmland and secondary forests are not:*** C. H. Mendenhall et al., "Predicting Biodiversity Change and Averting Collapse in Agricultural Landscapes." *Nature* 509 (May 8, 2014): 213–217. www.nature.com/nature/journal/v509/n7499/full/nature13139.html.

253 ***"almost one-half of the common bat species":*** Marty Downs, "Island Biogeography Theory Misses Mark for Tropical Forest Remnants," Nature Conservancy blog, April 16, 2014. blog.nature.org/science/2014/04/16/island-biogeography-theory-forest-remnants/.

253 ***a theory of "countryside biogeography":*** Henrique M. Pereira and Gretchen C. Daily, "Modeling Biodiversity Dynamics in Countryside Landscapes." *Ecology* 87.8 (August 2006): 1877–1885. www.azoresbioportal.angra.uac.pt/files/publicacoes_PEREIRA06_Biodiversity Countryside.pdf.pdf.

253 ***"Nature is almost everywhere":*** Emma Marris, *Rambunctious Garden: Saving Nature in a Post-Wild World.* New York: Bloomsbury, 2011, 224.

254 ***anthromes are mosaics of land:*** Michael P. Perring and Erle C. Ellis, Chapter 8, "The Extent of Novel Ecosystems: Long in Time and Broad in Space," in *Novel Ecosystems: Intervening in the New Ecological World Order,* Richard J. Hobbs, Eric S. Higgs, and Carol M. Hall, eds. New York: Wiley, 2013, 66–80. ecotope.org/people/ellis/papers/perring_2013.pdf.

254 ***Earth is an extensively modified used planet:*** Erle C. Ellis et al., "Used Planet: A Global History," *Proceedings of the National Academy*

of Sciences 110.20 (May 14, 2013): 7978–7985. ecotope.org/people/ellis/
papers/ellis_2013b.pdf.

255 ***"Imagine that an alien scientist":*** James H. Brown and Dov F. Sax, "Bio-
logical Invasions and Scientific Objectivity: Reply to Cassey et al." *Austral
Ecology* 30 (June 2005): 481–483. www.brown.edu/Research/Sax
_Research_Lab/Documents/PDFs/Biological%20invasions.pdf.

256 ***"biologists cannot tell by observation":*** Mark Sagoff, "What Does
Environmental Protection Protect?" *Ethics, Policy and Enviroment* 16.3
(December 2, 2013): 239–257. www.tandfonline.com/doi/full/10.1080/
21550085.2013.843362#U59TQShWj_Y.

256 ***directional and deterministic process of succession:*** Steward T. A.
Pickett and J. M. Grove, "Urban Ecosystems: What Would Tansley Do?"
Urban Ecosystems 12 (January 20, 2009): 1–8.

256 ***"theory of the climax state":*** Robert Nelson, "Ecological Science as a
Creation Story." *Independent Review,* Spring 2010, 513–534.

257 ***stable interdependent communities as the norm:*** Dr. Arthur Shapiro,
"Composition of Ecological Communities Is Dynamic." Commonwealth
Club lecture, May 8, 2014. milliontrees.me/2014/05/08/dr-arthur
-shapiro-composition-of-ecological-communities-is-dynamic/.

257 ***"ecologists have come to understand the reality":*** John Kricher,
"Nothing Endures but Change: Ecology's Newly Emerging Paradigm."
Northeastern Naturalist 5.2 (1998): 165–174. biophilosophy.ca/Teaching/
2070papers/kricher.pdf.

257 ***Ecological fitting is the process:*** Salvatore J. Agosta and Jeffrey A.
Klemens, "Ecological Fitting by Phenotypically Flexible Genotypes:
Implications for Species Associations, Community Assembly and Evolution." *Ecology Letters* 11.11 (November 2008): 1123–1134. onlinelibrary
.wiley.com/doi/10.1111/j.1461-0248.2008.01237.x/abstract;jsessionid=A4
31ABA8A6A229AFA3B54DE9747AD57D.f01t01.

258 ***Species don't need to coevolve:*** David M. Wilkinson, "The Parable of
Green Mountain: Ascension Island, Ecosystem Construction and Ecological Fitting." *Journal of Biogeography* 31 (January 2004): 1–4. www
.staff.livjm.ac.uk/biedwilk/pdfs/greenmt.pdf.

258 ***"the popular view [is] that diversity":*** Dov F. Sax and Steven D. Gaines,
"Species Diversity: From Global Decreases to Local Increases." *Trends in*

Ecology and Evolution 18.11 (November 2003): 561–566. planet.uwc.ac.za/nisl/Invasives/Refs/SaxandGaines.pdf.

259 *"North America presently has more":* James H. Brown and Dov F. Sax et al., "Aliens Among Us." *Conservation,* July 29, 2008. conservation magazine.org/2008/07/aliens-among-us/.

258 *analyzed a massive data set:* Maria Dornelas et al., "Assemblage Time Series Reveal Biodiversity Change but Not Systematic Loss." *Science* 344.6181 (April 18, 2014): 296–299. www.sciencemag.org/content/344/6181/296.short.

259 *New Zealand's 2,000 native plant species:* Dov F. Sax, Steven D. Gaines, and James H. Brown, "Species Invasions Exceed Extinctions on Islands Worldwide: A Comparative Study of Plants and Birds." *The American Naturalist,* December 2002, 766–783. labs.bio.unc.edu/Peet/courses/bio255_2003f/papers/Sax2002.pdf.

259 *In California, an additional 1,000 new species:* James H. Brown and Dov F. Sax, "An Essay on Some Topics Concerning Invasive Species." *Austral Ecology* 29 (2004): 530–536. www.bio.fsu.edu/miller/docs/Brown_SAX.pdf.

260 *overall species richness of the plant life on Pacific islands:* Dov F. Sax and Steven D. Gaines, "Species Invasions and Extinction: The Future of Native Biodiversity on Islands." *Proceedings of the National Academy of Sciences,* Supplement 1, August 12, 2008, 11490–11497. www.pnas.org/content/105/Supplement_1/11490.short.

260 *introduced species of plants and animals:* Joan G. Ehrenfeld, "Ecosystem Consequences of Biological Invasions." *Annual Review of Ecology, Evolution, and Systematics* 41 (December 2010): 59–80. izt.ciens .ucv.ve/ecologia/Archivos/ECO_POB%202010/ECOPO4_2010/Ehrenfeld%202010.pdf.

260 *"meta-analysis of over 1000 field studies":* Barry W. Brook et al.,"Does the Terrestrial Biosphere Have Planetary Tipping Points?" *Trends in Ecology and Evolution,* July 2013, 396–401. www.sciencedirect.com/science/article/pii/S0169534713000335.

260 *"In truth, ecologists and conservationists":* Martin Jenkins, "Prospects for Biodiversity," *Science* 302.5648 (November 14, 2003): 1175–1177. www .zo.utexas.edu/courses/Thoc/Readings/Jenkins_Science2003.pdf.

260 *Pleistocene Rewilding proposal:* Josh Donlan et al., "Re-Wilding North America." *Nature,* August 18, 2005, 913–914. izt.ciens.ucv.ve/ecologia/Archivos/ECO_POB%202010/ECOPO4_2010/Ehrenfeld%202010.pdf.

261 *"Paleolithic landscape at the Oostvaardersplassen":* Sagoff, "What Does Environmental Protection Protect?" *Ethics, Policy, & Environment,* 16, No. 3, (2013): 239–257.

262 *"ecosystems have no preferences":* Robert T. Lackey, "Values, Policy, and Ecosystem Health." *BioScience* 51.6 (June 2001): 437–443. fw.oregonstate.edu/system/files/u2937/2001c%20-%20Values,%20Policy,%20and%20Ecosystem%20Health%20-%20Reprint%20-%20Lackey.pdf.

262 *"to an agronomist":* Peter Kareiva et al., "Domesticated Nature: Shaping Landscapes and Ecosystems for Human Welfare." *Science* 316.5833 (June 29, 2007): 1866–1869. faculty.washington.edu/timbillo/Readings%20and%20documents/ABRIDGED%20READINGS%20for%20PERU/kareiva_etal_2007.pdf.

262 *"Humans must proactively manage ecosystems":* Kricher, "Nothing Endures but Change: Ecology's Newly Emerging Paradigm." *Northeastern Naturalist,* Vol.5, No. 2 (1988), 165–174, biophilosophy.ca/Teaching/2070papers/kricher.pdf.

INDEX

attention deficit hyperactivity disorder (ADHD), 121–22

Ausubel, Jesse, 61, 65, 66, 249–50

autonomous vehicles, 63–64

Bacillus thuringiensis (Bt), 134–35, 148, 164–65

Bailey, Ronald
    cancer concern of, 95–96
    climate change work and view of, 171–72, 173, 232
    energy sector work of, 31–32
    works by, xiii, xv–xvi, xix

Ban Ki-moon, 67

Bangladesh, 19, 29

Barnosky, Anthony, 240

Barthel, Fabian, 198–99

Bastiat, Frédéric, 86

Benbrook, Charles, 144–45

Berg, Paul, 130–31

biases, cognitive, xvii–xviii, 80, 109, 123, 220–22

biodiversity. *See* extinction; nature

biofuels, 82, 168

biology, synthetic, 66, 85

biotech crops
    accidental pollination by, 158–60
    benefits overview, 265
    blindness prevention by, 152
    developing nations' use of, 137–38, 149–54, 164–65, 168
    doomsayers, 79, 130, 133–34, 135–38, 141–48
    economic gains from, 135, 146
    environmental benefits of, 66, 135, 148–49, 150, 166
    Europe's position on, 138, 139–40, 141–43
    farmer suicide myth and, 164–65
    fertilizer nutrient efficiency of, 56, 167

    future innovations of, 165–69
    India's concern about, 136
    introduction of, 134–35, 151
    labeling, 160–63
    lab-grown meat, 66
    lawsuit cases by, 158–60
    mutation concerns for, 146–47
    organics and, 156–58, 161–63, 170
    pesticides and, 134–35, 143–46
    safety of, 135, 139–41, 169
    superpests and, 154–55
    superweeds and, 155–56
    tolerance of, 133, 166–67

biotechnology
    animals, 168–69
    health, 85, 88, 130–33
    trees, 166–67

birds, 101, 109–13, 239–40

birth control pills, 122

birth defects, 119–20

Bitman, Joel, 111

Bittman, Mark, 141

blindness, 152

Bloch, Harry, 59

Bonny, Sylvie, 136

Borlaug, Norman, 10–13

Bouwer, Laurens, 198

bovine somatotropin (BST), 134

Bowman, Vernon Hugh, 159–60

BP, 42, 45

Bradley, Raymond, 69

Brazil, 19, 203

Breakthrough Institute report, 234–35

breast
    cancer, 106, 113–15, 118
    development, 121

Bretschger, Lucas, 72–73

Breyman, Steve, 78

Brin, Sergey, 66

Brodeur, Paul, 83

Brookes, Graham, 144

Brown, James, 255, 259
Brown, Lester, xvi, 4, 36
Brown, Patrick, 184
Brundtland, Gro Harlem, 67
Brunk, Gregory, 70, 71–72
BST. *See* bovine somatotropin
Bt. *See Bacillus thuringiensis*
Bulled, Nicola, 21
Burbach, David, 28

Caldeira, Ken, 213, 235
Campbell, Colin, 40
cancer
    aging and, 104–5, 115–17
    Bailey's personal concern for, 95–96
    breast, 106, 113–15, 118
    cell phones and, 84
    DDT and, 102–6, 113–15
    doomsayers, 96–97, 102–6, 114–15
    EMFs and, 83–84, 123
    endocrine disruption and, 118, 120
    false positives for, 123
    incidence decrease in, 97–98, 106–7,
      117, 265
    incidence increase in, 96–97, 115–17
    leukemia, 83, 104, 123
    lung, 95–96, 103, 117
    mortality rate, 97–98, 104, 116
    pancreatic, 123
    pharmaceuticals, 81–82
    pollution and, 96, 106–7
    risk factors, 107
    saccharin and, 82–83
    synthetic chemicals and, 96–97, 105,
      106–7, 114
    testicular, 120
    tobacco smoking and, 95–96
capitalism, free-market, 67–68, 71–72
carbon emissions
    budget and projections, 176–78
    capture and storage, 215

climate sensitivity to, 187–90
consensus on, 232–34
current, 175
Kyoto Protocol on, 201, 206, 209,
    232–33
natural gas reduction in, 235
nuclear power reduction in,
    222–25
ocean acidification from, 190–91
renewable energy and, 226–37
tax, 208–10
trade, 45–46, 201, 206–8, 209,
    232–34
Carson, Rachel
    on DDT, 99, 101–6, 113
    modern environmentalism from, 98,
      108–9
    on synthetic estrogen, 117
Carter, Jimmy, 31, 33
Cavalieri, Liebe, 131
CBD. *See* Center for Biological
    Diversity
CDC. *See* Centers for Disease Control
    and Prevention
cell phones, 62–63, 64, 84
Center for Biological Diversity (CBD),
    xvi, 240, 243
Centers for Disease Control and
    Prevention (CDC), 101, 106
CFCs. *See* chlorofluorocarbon
chemicals, synthetic, 96–97, 105, 106–7,
    114. *See also* DDT; endocrine
    disrupting chemicals
China
    climate negotiations with, 202–4
    neodymium from, 57–58
    sulfur dioxide emissions in, 60
chlorofluorocarbon (CFCs), xvi
Christy, John, 180
Clayton, Blake, 38
Clements, Frederic, 256

politicization of, 109

precautionary principle positioned for, 75–77, 87, 136

reproductive problems, 101, 109

saccharin and, 82–83

sperm, 117–19

synthetic biology for, 85

Heinberg, Richard, xvi, 35

herbicides, 135, 143–45, 148, 155–56

Heritage Foundation, 26

Hickey, Joseph, 111

HIV/AIDS, 19–20

Holdren, John, 34, 39

homeopathy, 126

Hooker, Joseph, 258

Hopfenberg, Russell, 9

hormones, in meat and dairy, 134. *See also* endocrine disrupting chemicals

Howard, Ted, 132–33

Hubbert, M. King, 41, 42

Hueper, Wilhelm, 102

hurricanes, 195–96, 197–98

hydraulic fracturing, 235

hypospadias, 119–20

IEA. *See* International Energy Agency

IIASA. *See* International Institute for Applied Systems Analysis

Iler, Stuart, 45

income increase

climate adaptation and, 192–94

climate mitigation and, 195

fertility rate decline and, 21–23

intergenerational equity and, 212–13

open-access social orders and, 71

trend overview, 264

India

biotech crops in, 137, 151, 164–65

climate change negotiations with, 203

farmer suicide in, 164–65

fertility rate and life expectancy in, 19

Green Revolution in, 10–12

oil consumption patterns for, 66

Orissa cyclone, 137

Industrial Revolution, 22, 91

industrialization

commodity super-cycles and, 37

fertility rate decline and, 22

innovation trial and error in, 91

pollution correlation to, 60–61

Information Technology and Innovation Foundation (ITIF), 226–29

Inhofe, James, 173

innovation

cognitive biases against, 80

elitist resistance to, 68–69, 89–90, 92–93

fertility rate decline and, 22

free-market capitalist drive for, 72

population projections and, 15–17, 30

positive possibilities with, 263–66

precautionary resistance to, 78–80, 84–86, 89–90, 92–93

trial and error for, 91–94

innovation sectors and types

additive manufacturing, 66–67

autonomous vehicles, 63–64

biofuel, 82, 168

biotech crops, 165–69

cellular, 62–63, 64

climate geoengineering, 214–15

DDT, 99–100

electric vehicle, 58

energy, clean, 236–37

energy efficiency, 61–64, 66–67, 222–25, 234–35

food production, 13–15, 66, 100–101, 165–69

lithium, 56–57
Lockheed Martin, 225
Lovejoy, Thomas, 242
Low, Bobbi, 18–19
lung cancer, 95–96, 103, 117
Lutz, Wolfgang, 16, 17, 24

MacArthur, Robert, 244
mad cow disease, 136
Maddison, Angus, 67, 212
malaria, 99–100, 113
male health, 117–20
Malthus, Thomas Robert, 5–7
manufacturing, additive, 66–67
Marris, Emma, 253, 254, 261–62
Marx, Karl, 248
Mascaro, Joe, 255
Maugeri, Leonardo, 43, 46–47
McCauley, Douglas, 246, 247
McKibben, Bill, 172
meat industry, 65, 66, 134, 136
metals
    commodity prices of, 34–35, 37, 38,
        57–58
    consumption plateau, 65–66
    reserve depletion, 52–54, 56–59
methane, 175
Mexico, 9, 10, 19
Miller, Georgia, 86
minerals
    commodity prices of, 34–35, 37, 38,
        57–58
    geological studies on, 39, 53, 55–56,
        59
Mokyr, Joel, 91, 92
monopolies, 68, 72
Monsanto, 14, 159–60
    Roundup, 155–56, 159
Montague, Peter, 77
Mooney, Chris, 77
moral restraint, 6

Morano, Marc, 173
mortality rates
    from cancer, 97–98, 104, 116
    coal and nuclear comparison of, 223
    fertility rates and, 2, 18–21
    from infectious disease, 116
    from natural disasters, 197
    water access and, 49
Motavalli, Jim, 41
Muir, John, 254
Müller, Paul, 99
Muñoz, Félix-Fernando, 15–16
Murtin, Fabrice, 23
Muufri, 169
Mwanawasa, Levy, 138
Myers, Norman, 241–42
Myhrvold, Nathan, 214

Naam, Ramez, 29–30, 61
nanotechnology, 85–86
NASA reports, 178, 189, 223, 224
National Academy of Sciences (NAS)
    on biotech crops, 140, 141, 146, 148
    on climate change mitigation, 214
    on DDT, 107
National Institute of Environmental
        Health Sciences, 124
National Oceanic and Atmospheric
        Administration (NOAA), 174, 186
natural disasters, 195–200
natural gas, 31–32, 45, 230, 231, 235
natural selection, 7
natural states theory, 68–70
nature
    doomsayer ideals of, 8, 98, 133,
        253–54, 256–57
    myth of balance in, 256–57
    myth of pristine nature, 253–56
    novel ecosystems, 255–56, 257–62
    restoration of, 248–53, 266
Nelson, Robert, 256–57

salmon, biotech, 169

Sanyal, Sanjeev, 16

Sapsford, David, 59

Saudi Arabia, 44, 47

Sax, Dov, 255, 259

Schindler, Jörg, 41

Schlesinger, James, 39, 41

Schmeiser, Percy, 159

science

    first ban on, 131

    literacy and polarization on, 219–20

    pathological, 124–27

    politicization of, 109

sea-level rise, 176, 177

Sedjo, Roger, 166

Seffrin, John, 97

Segerfeldt, Fredrik, 50–52

self-driving vehicles, 63–64

Séralini, Gilles-Éric, 141–42

Shapiro, Arthur, 257

Shapiro, Samuel, 124

Shiva, Vandana, 137, 158, 164

Silent Spring (Carson), 98–99, 101–5, 107, 108–9, 113, 117

Simmons, Matthew, 40

Simon, Julian, 34, 36

Singer, Fred, 173

Singer, Hans, 36

Smil, Vaclav, xv, 61–63

smoking, cancer and, 95–96

Smulders, Sjak, 72–73

societal collapse, 68–71

solar power, 227, 228, 229–32

Sosis, Richard, 21

Souder, William, 103, 108

soybeans, 14, 56, 137, 147, 149

sperm quality and count decline, 117–19

Speth, James Gustave, 222, 223

SPICE. *See* Stratospheric Particle Injection for Climate Engineering

storms, 195–96, 197–98

Stratospheric Particle Injection for Climate Engineering (SPICE), 215

subsidies

    cutting, 210–11

    energy, 32, 211, 236–37

    food production, 11–12, 50, 211

Sugiyama, Taishi, 232–33

sulfur dioxide, 60–61, 182, 214

Sunstein, Cass, 78, 80

superpests, 154–55

superweeds, 155–56

sustainable development, 67–73. *See also* climate change mitigation

Swanson, Richard, 230

Sweeney, Edmund, 107–8

Synfuels Corporation, 31–32

synthetic biology, 66, 85

synthetic chemicals, 96–97, 105, 106–7, 114. *See also* DDT; endocrine disrupting chemicals

Tahil, William, 56

Tainter, Joseph, 69–70

technology. *See* biotech crops; biotechnology; innovation

temperature increase

    climate sensitivity and, 187–90

    projections, 176–77, 180–81, 186

    trends, 174, 175–76, 179

temperature increase hiatus

    modeling error and, 179–81, 183–84, 185, 187

    state of, 178–79

    theories on, 181–83, 185, 186

TerraPower, 224–25

Tesla, 56–57, 58

testicular cancer, 120

Thierer, Adam, 78

thorium, 223–24